Feed Water Sys
and Treatment

Feed Water Systems and Treatment

G.T.H. Flanagan, CEng, FIMarE, MRINA

Newnes
An imprint of Butterworth-Heinemann Ltd
Linacre House, Jordan Hill, Oxford OX2 8DP

PART OF REED INTERNATIONAL BOOKS

OXFORD LONDON BOSTON
MUNICH NEW DELHI SINGAPORE SYDNEY
TOKYO TORONTO WELLINGTON

First published by Stanford Maritime Ltd 1978
Revised and reprinted 1983
Reprinted 1987
Reprinted by Butterworth-Heinemann Ltd 1991

ISBN 0 7506 0368 2

Printed and bound in Great Britain by
Biddles Ltd, Guildford and King's Lynn

CONTENTS

Preface

The aim of this book is to provide basic information on feed water systems and on chemical treatment for feed and boiler water.

Although aimed at the requirements of the Second Class Certificate of Competency for Marine Engineer Officers, it should also provide useful revision for those engineers preparing for their First Class Certificate.

The diagrams in this book are not drawn to scale and show the type of simplified sketch required for examination purposes, although there will not be sufficient time during the examination to attain the same level of neatness, the Candidate attempting to do as much as possible of the sketch freehand.

A list of the SI Units used in this book is given together with some conversions from other systems of units.

G. T. H. FLANAGAN

SI UNITS

Mass = kilogramme (kg)
Force = newton (N)
Length = metre (m)
Pressure = newton/sq metre (N/m^2)
Temperature = degree Celsius (°C)

CONVERSIONS

1 inch = 25·4 mm = 0·025 m
1 foot = 0·3048 m
1 square foot = 0.093 m^2
1 cubic foot = 0.028 m^3
1 pound mass (lb) = 0.453 kg
1 UK ton (mass) = 2240 lb = 1016 kg
1 short ton (mass) = 2000 lb = 907 kg
1 tonne mass = 1000 kg
1 pound force (lbf) = 4·45 N
1 ton force (tonf) = 9·96 kN
1 kg = 9.81 N
0·001 in = 0·025 mm
1 lbf/in^2 = $6895N/m^2$ = $6·895kN/m^2$
1 kg/cm^2 = 1 kp/cm^2 = $102kN/m^2$
1 atmos = 14·7 lbf/in^2 = $101.35kN/m^2$ = 760 mm Hg
1 bar = 14·5 lbf/in^2 = 100 kN/m^2

Note: For approximate conversion of pressure units

100 kN/m^2 = 1 bar = 1 kg/cm^2 = 1 atmos
1 $tonf/in^2$ = 15440 kN/m^2 = 15·44 MN/m^2
(°F—32) x $\frac{5}{9}$ = °C
1 HP = 0·746 kW
1 ppm = 1 gm dissolved solids/ 1 000 000 gm pure water
1 Thirty-second = 5 oz dissolved solids/160 oz pure water
$\frac{1}{32}$= 2200 grains/gallon = 32000 ppm
1 ml/litre dissolved oxygen x 1·5 = ppm of dissolved oxygen

CHAPTER 1

Feed Systems

To maintain the quantity of water required for the boiler, marine feed systems condense steam, their basic function being to collect and condense exhaust steam from main and/or auxiliary plant and then to return the condensed water, referred to as condensate, to the boiler.

In its simplest form the system consists of a condenser which removes the latent heat from the exhaust steam, so that the resulting condensate can enter a drain tank from which it is delivered to the boiler by means of a feed pump. As the drain tank, often referred to as the hot well, is open to the atmosphere, this arrangement is known as an open feed system.

To allow for leakage and other losses, a certain amount of make up feed will be necessary. This should consist of fresh water, obtained either by evaporating sea water or from shore supplies, and stored ready for use in a suitable reserve feed tank.

While a condenser working under atmospheric conditions is suitable for simple auxiliary plant, for more involved systems vacuum conditions are necessary in the condenser so that the steam can be expanded to lower exhaust pressures, the resulting lower temperatures now obtained increasing the thermal efficiency of the plant. These vacuum conditions require a pump to be fitted in order to remove the condensate from the low pressure existing in the condenser to the atmospheric pressure at the hot well. In relatively simple systems where only modest vacuum is produced, a single air pump can be fitted to remove the condensate, together with air and vapour from the condenser sump. Where higher vacuum conditions are involved, two pumps will be required. The condensate will be removed by an extraction pump while a separate air pump will use a separate take off point, remote from and higher than the condensate outlet, to extract the air and vapour.

On contact with the atmosphere water will take oxygen into solution and this can lead to conditions where corrosion may take place. The rate of any corrosive action thus produced increases with higher temperatures and this can cause serious corrosion problems especially in high performance water tube boilers.

To remove the bulk of the oxygen and other dissolved gases from the water in systems operating at higher pressures a deaerator may be included in the system. In this unit the water is raised to its boiling point, and the majority of the dissolved gases are released and can then be removed before the feed water enters the boiler. At even higher operating pressures this absorption of oxygen is greatly reduced by preventing the feed water from coming into contact with the atmosphere by means of a closed feed system, such as is used with turbine installations.

A problem which arises in all closed feed systems is the differing amounts of water actually required in circuit with varying boiler load conditions. At high load a large number of steam bubbles form below the water level in the boiler, thus reducing the mass of water it contains if a constant water level is maintained. This leads to an excess of water in the feed circuit and some must be removed to a suitable storage tank. When the boiler load is then reduced, there is a corresponding decrease in the formation of steam bubbles below the water level in the boiler, and to maintain a constant water level more water must be supplied. Thus means must be provided to rapidly return the water from the storage tank into the main feed circuit. This is considered in some of the following systems.

The arrangement of individual feed systems varies greatly from one ship to another and only a number of representative layouts will be considered here, each chosen to illustrate specific points.

OPEN FEED SYSTEM

The layout shown in Fig. 1 is suitable for a relatively simple auxiliary plant.

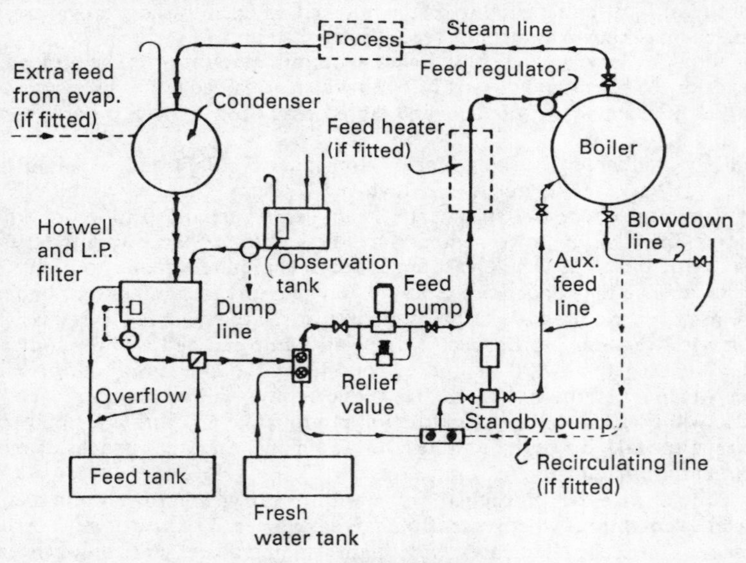

Fig. 1 Open Feed System

An auxiliary condenser, working at atmospheric pressure and using sea water as a coolant, is used to condense the exhaust steam and to cool hot water drains. The resultant condensate drains by gravity into a low pressure feed filter which also serves as the hot well as shown in Fig. 2.

Any hot water drains which could be oil contaminated are led into a suitable observation tank (see Fig. 3) before entering the feed circuit.

The feed pump normally draws water from the hot well, which can overflow into a reserve feed tank. When the level in the hot well falls below a predetermined point, a float operated change-over valve allows the feed pump to draw directly from the reserve feed tank. The pump shown is an electrically driven displacement type running at constant speed. The water in the boiler is main-

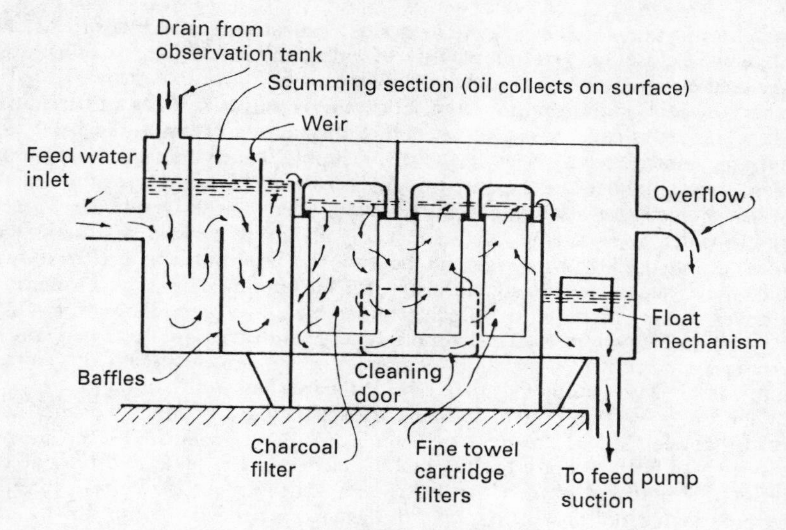

Fig. 2 Low Pressure Feed Filter

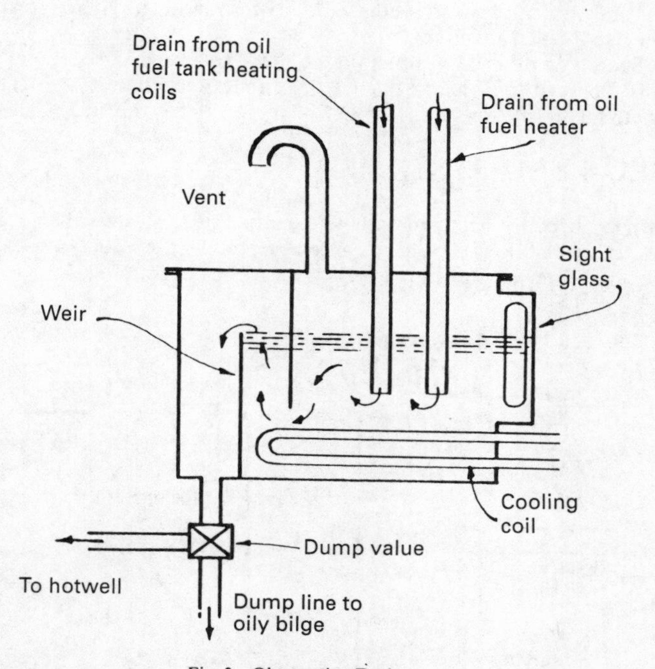

Fig. 3 Observation Tank

tained at constant level by means of a float operated feed regulator. As this control valve closes in to restrict the flow of water, any undue increase of pressure at the feed pump discharge is prevented by means of a spring loaded relief valve discharging back to the pump suction. For standby purposes, or as an alternative arrangement, a steam driven direct acting displacement pump can be fitted. A suitable control valve placed in the steam supply line enables this pump to be driven at variable speed in response to signals from the boiler feed regulator.

When the system is used in conjunction with a small shell type boiler where a variable water level can be tolerated, a less elaborate on and off method of control can be used. In this, when the boiler water level falls to a predetermined value, a level sensor sends a signal to start the feed pump. This pumps water into the boiler, raising the level to a predetermined high level when the sensor causes the pump to be switched off. A simple float-operated magnetic switch can be used to give the control signal and in conjunction with an electrically driven feed pump would be very suitable for the type of feed system shown in Fig. 1.

When an open feed system is being used to supply a boiler working at a pressure above 1800 kN/m^2 it is recommended that a deaerator be included in the system to both increase the thermal efficiency of the plant and remove the bulk of the dissolved oxygen. An alternative arrangement is to fit an internal feed pipe in the boiler which sprays the water entering the drum downwards through the steam space so allowing the boiler to act as its own deaerator, the released air passing out with the steam. While this arrangement protects the boiler, the air is not removed from the system as is the case with a proper deaerator and this can lead to increased corrosion in the auxiliary steam lines in the presence of wet steam conditions.

When Scotch or other large shell type boilers are fitted, a recirculating line is commonly included to give improved water circulation in the boiler when raising steam. With water tube or smaller shell boilers this line is omitted.

CLOSED FEED SYSTEMS

These can be broadly grouped into two main types, those suitable for boiler pressures up to 4000 kN/m^2 and those for pressures above 3000 kN/m^2. A basic layout for the former is shown in Fig. 4 and for the latter in Fig. 7.

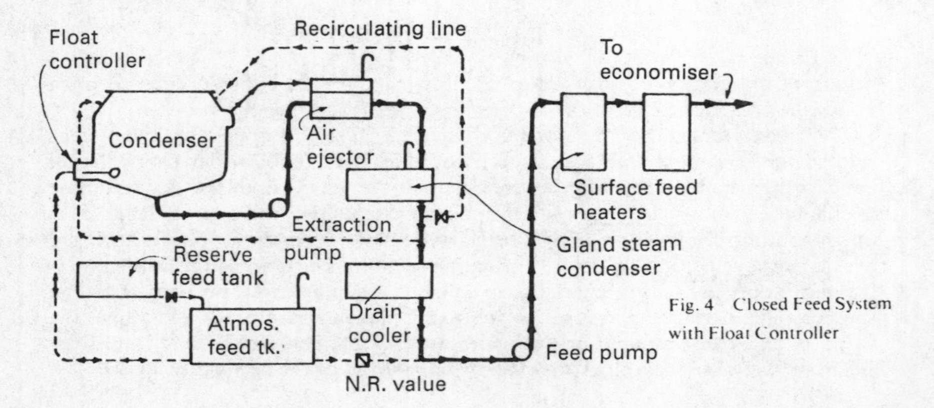

Fig. 4 Closed Feed System with Float Controller

The system shown in Fig. 4 starts with a regenerative type condenser receiving the exhaust steam from the main turbines. The resulting condensate is then removed by an extraction pump. The necessary suction head for this pump being maintained by keeping the condensate level approximately constant in the condenser sump. This is achieved by means of a float operated controller mounted either in the condenser, or adjacent to it in a suitable chamber with balance connections to the condenser.

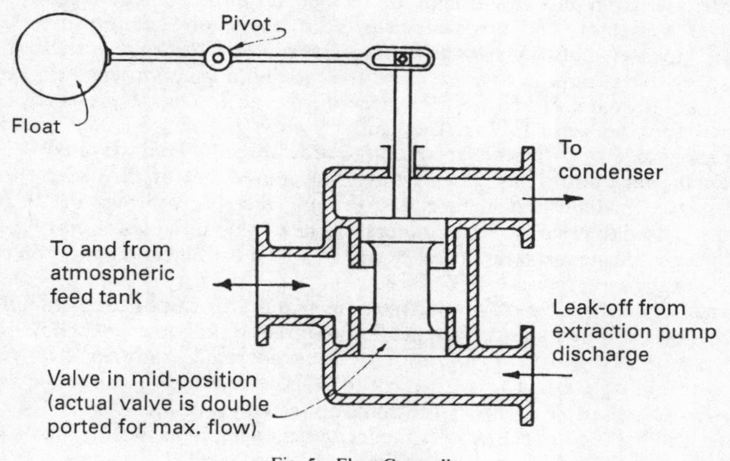

Fig. 5 Float Controller

As shown in Fig. 5 the float operates a balanced valve which directs a leak off from the extraction pump discharge to the atmospheric feed tank, or from this tank into the main condenser. The valve has a neutral position about its mid travel so that it does not operate for slight variations from the normal operating level. To increase the flow rate without undue valve travel in actual practice the valve is double ported.

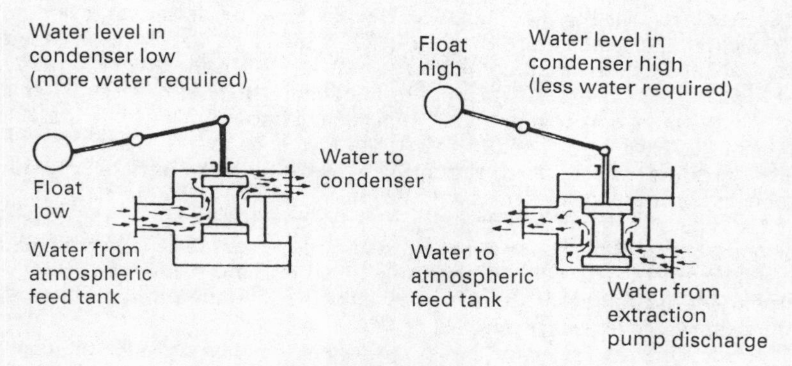

Fig. 6 Operation of Float Controller

Operation of the controller is indicated in Fig. 6. A high condensate level in the condenser sump is reduced by pumping water via the controller into the atmospheric feed tank, whereas for a low level the valve assumes a position where atmospheric pressure causes water from this tank to flow into the main condenser. In the controller shown, the valve is directly operated by the float, but in some of the larger units the float only works a small relay valve which operates an actuator for the main control valve.

Feed water from the extraction pump then passes through an air ejector, where it acts as a coolant, and flows on to perform the same function in the gland steam condenser and drain cooler before entering the feed pump suction. The atmospheric feed tank is also linked to the feed pump suction via a non return valve normally kept closed by the extraction pump discharge pressure. In the event of such pressure failing, the head of water in the feed tank opens this emergency valve allowing the feed pump to draw directly from the tank.

After the feed pump has raised its pressure above that of the boiler the feed water passes through two surface feed heaters. Here, steam bled off from the turbines is used to raise the feed temperature to a value suitable for entry into the economiser. This temperature must exceed 115°C if fouling and corrosion on the gas side of the economiser is to be avoided.

Extra feed in the form of vapour from an evaporator can be supplied either to the shell of the first stage surface heater or to the drain cooler, depending upon the pressure at which the evaporator is being operated. Another arrangement used with multiple effect or with flash off type evaporators, where the vapour has been condensed, is to supply the make up feed into the feed tank.

After leaving the surface heaters the feed water usually passes through a boiler feed water regulator before entering the economiser.

The various drains and lines to supply heating steam to the various heat exchangers have been omitted from the diagram for the sake of clarity, but the important recirculating line is shown. This is fitted so that at low load conditions or when manoeuvring, sufficient water continues to be circulated through the steam jet air ejector and gland steam condenser to prevent these overheating even when the boiler feed water regulators are closed.

A disadvantage with this type of feed system is that using the atmospheric feed tank as a surge tank in the manner described, creates a tendency for the water in it to become aerated, so raising the dissolved oxygen content of the feed when this water is returned to the condenser. Heating the water in the surge tank to a sufficiently high temperature to drive off this dissolved oxygen would be impracticable as well as resulting in a large heat loss when the boiling water was returned to the condenser. Attempts to maintain a blanket of steam over the water in the surge tank to prevent air from coming into contact with it, and other similar ideas have been found in practice to give poor results. Thus for higher boiler pressures where dissolved oxygen leads to more serious corrosion problems the alternative arrangement shown in Fig. 7 is usually preferred.

This system again starts with a regenerative condenser, but in this case the condensate level in the condenser sump is allowed to vary. The extraction pump must therefore be of a special design, able to self regulate its rate of discharge to suit varying suction heads, it does so by allowing cavitation to occur at a controlled rate in the first stage impeller.

After leaving the extraction pump the feed water then provides the cooling medium as it circulates through the air ejector, gland steam condenser and drain

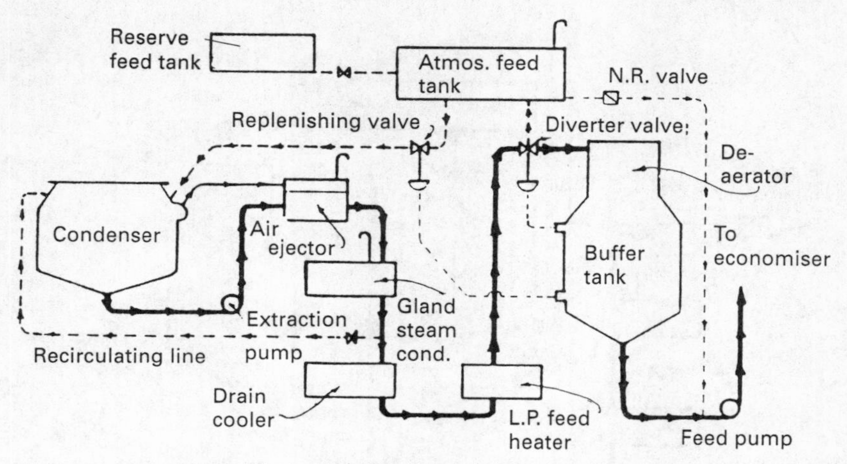

Fig. 7 Closed Feed System with Deaerator

cooler. To raise the water to a temperature suitable for entry into the deaerator a low pressure surface heater is included in the system. The feed water then passes through a diverter valve to enter the deaerator. After being deaerated, it passes into a storage tank mounted directly beneath the deaerator. This also acts as a buffer or surge tank where the water can be stored in a hot, deaerated condition ready for immediate use. When the level in this tank rises above a predetermined value, the diverter valve allows a proportion of the feed water to enter the atmospheric feed tank in which the excess water can be stored until required. When the level in the buffer tank falls below a predetermined low level, the replenish valve opens allowing water from the atmospheric feed tank to enter the main condenser. Under normal operating conditions the buffer tank holds sufficient water to cope with reasonable changes of load, and no transfer of water to or from the atmospheric feed tank will be required, thus less aerated water enters the feed circuit helping to maintain a low dissolved oxygen content in the system.

After leaving the buffer tank the water enters the feed pump where its pressure is raised high enough for it to be delivered into the boiler as required. Deaerators use contact heat exchange which means the water leaving them will be close to its saturation temperature. Slight pressure reduction at the feed pump inlet can lead to some of this water flashing off as steam which can cause the pump to gas up and lose its suction. This may be avoided by increasing the pressure and raising the corresponding saturation temperature above the actual temperature of the water, so preventing such formation of vapour. The usual method of doing this is to mount the deaerator and its buffer tank high in the engine room to increase the head of water above the suction. An alternative sometimes used, is to fit a boost pump at the discharge from the buffer tank.

If the feed temperature leaving the feed pump is now above 115°C it can enter the economiser without further heating. However, if below this temperature, or if higher feed temperatures are desired to increase the thermal efficiency of the plant, additional surface heaters using steam bled from the turbines as the heating medium, can be fitted after the feed pump.

In tankers and other vessels with large auxiliary steam loads in port, this basic system may be modified as shown in Fig. 8.

A drain tank is now included in the system, to which drains from various

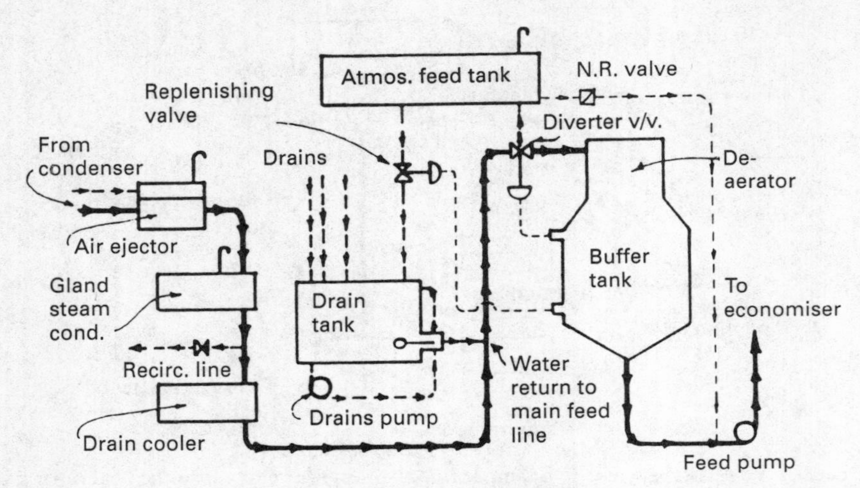

Fig. 8 Closed Feed System with Drains Tank

auxiliary units can be led. This tank is fitted with its own extraction pump which, together with a float controller similar to that previously described for a main condenser, keeps the water level in the tank approximately constant. If this level tends to rise the excess water is pumped via the controller into the main feed circuit as shown. A high level in the buffer tank is dealt with as in the previous arrangement, a proportion of the feed being diverted to the atmospheric feed tank, while a low level causes the replenish valve to open. This now supplies water to the drain tank where, to keep the level constant, the float controller allows more water to be pumped into the main circuit so increasing the feed flow to the deaerator and restoring the level in the buffer tank. The advantage of this arrangement is that the deaerator can still be used when the main condenser is shut down.

PACKAGE FEED SYSTEMS

This refers to the practice of placing several units together in one casing or on a common bedplate. The resulting package can thus be assembled at the manufacturers workshops and delivered to the ship as a complete item. This procedure can be used for various parts of the system, for example a unit combining the air ejector, gland steam condenser and drain cooler, or in more extreme form applied to the whole feed system. This is usually only done in the case of small to medium sized auxiliary installations but can also be used for larger or main systems.

The main advantage claimed is that all work can be carried out to the same level of specification, and correct pipe sizes and valves used. The main disadvantage lies in the fact that to keep the overall dimensions of the package reasonably small for ease of transportation and handling, access to various fittings may be limited, leading to difficulties when carrying out subsequent maintenance work. A representative layout for a package feed system is shown in Fig. 9.

The extraction pump receives condensate from the condenser and then pumps

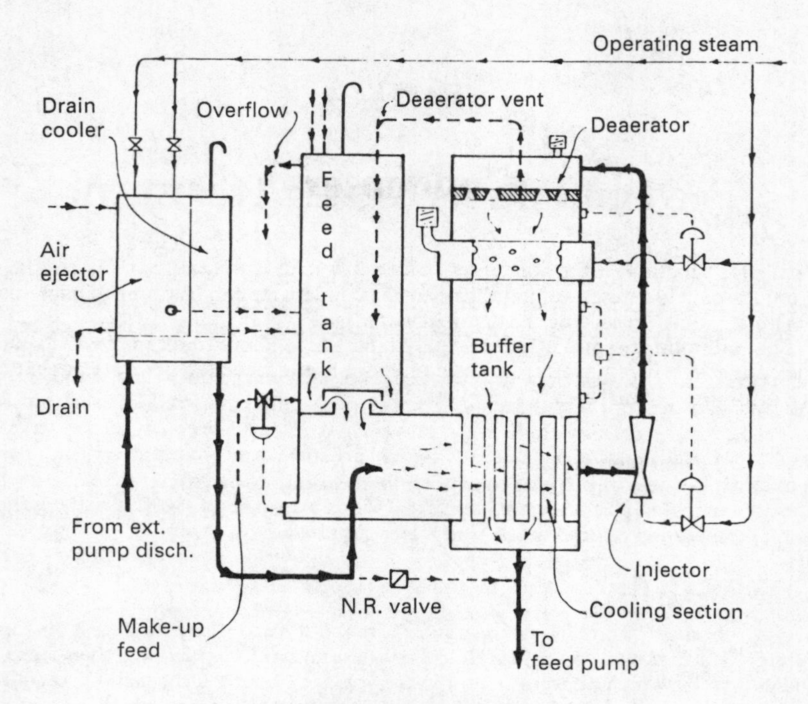

Fig. 9 Closed Feed System, Weir's Package Type

it through the air ejector and gland steam condenser, where it acts as the cooling medium, before entering the lower part of the atmospheric feed tank. A steam operated injector then supplies this water to the deaerator. The operating steam from the injector nozzle mixes with this feed helping to raise its temperature to such a point that, when sprayed through the deaerator nozzles, some of its mass immediately flashes off into vapour. This assists the deaerating process to such an extent, that the vent condenser normally associated with a deaerator can be omitted, any escaping vapour being condensed by placing the vent outlet below the water level in the feed tank. Heating steam supplied to the mixing chamber in the normal manner provides for further removal of dissolved gases before the deaerated water finally enters the buffer tank. To permit this to be placed adjacent to the feed pump, feed water on its way to the deaerator is circulated through a heat exchange section in the buffer tank. The water stored in this tank is thus cooled sufficiently to be delivered direct to the feed pump without risk of vapour forming at the pump suction.

CHAPTER 2

Condensers

The basic function of the condenser in a feed system is to remove the latent heat from the exhaust steam so that the resulting condensate can be pumped back into the boiler, thus conserving the distilled water feed.

For optimum thermal efficiency only latent heat should be removed, so that the condensate is removed from the condenser at the same temperature as that of the incoming exhaust steam.

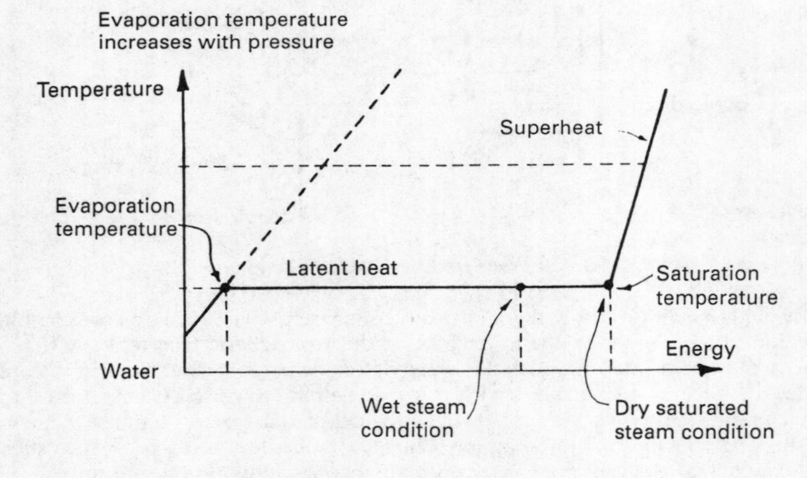

Fig. 10

The reason for this is that any energy given to the condenser cooling water is discharged at the ships side and so lost to the system, the energy then having to be restored by the combustion of fuel for evaporation back into steam to take place. The process which can cause an additional energy loss is that referred to as undercooling. This means that the condensate has been cooled below its corresponding evaporation temperature, and as seen in Fig. 10 this energy has to be restored before evaporation can again begin to take place. This loss is not of great importance in simple auxiliary condensers, but can lead to severe losses in condensers associated with turbine plant.

Another important function of the condenser is connected with the fact that when the exhaust steam is condensed, its volume is enormously reduced, by a factor in the order of 20 000 to 1. This enables a constant low pressure to be maintained in the condenser into which the exhaust steam will flow. If the process is carried out rapidly, while at the same time any non-condensable gases such as air are removed, pressures well below atmospheric pressure, referred to as vacuum conditions, can be produced. This low pressure enables the exhaust steam to be expanded down to a corresponding low saturation temperature, thus increasing the thermal efficiency of the plant.

It also follows that where high vacuum conditions are to be maintained it becomes an important consideration that the condenser also acts as a deaerator to give increased efficiency to the removal of the air and other non-condensable gases. There are a number of reasons why this must be done, the most critical being that the specific volume of these gases remains relatively large and if not removed would cause a back pressure to build up in the condenser leading to an increased exhaust steam temperature, so reducing the thermal efficiency of the plant. Air is also a very poor conductor of heat, and pockets and films of air forming around the tubes greatly reduce the rate of heat transfer between the exhaust steam and the cooling water. The presence of air also increases the amount of undercooling. This is due to the fact that assuming constant temperatures, the total pressure exerted by a mixture of gases equals the sum of the constituent pressures, each gas exerting such pressure as it would exert if it alone occupied the space. Now assuming the exhaust steam to behave as a gas, at the top of the condenser it is responsible for virtually all the pressure, the effect of the air being so small that it can be neglected. However, in the lower portion of the condenser, most of the steam will have been condensed, whereas the mass of air not being affected by the condensation process will have remained constant and so now forms a considerable proportion of the gas mixture. Hence although the total pressure remains approximately constant from top to bottom of the condenser, the fraction of this pressure due to the steam is less at the bottom than at the top if appreciable amounts of air are present. This means that in the lower parts of the condenser the vapour will have its saturation temperature reduced by a corresponding amount, and will have to be cooled to a lower temperature before it finally condenses. Any condensate coming into contact with this vapour will have its temperature reduced below that of the saturation temperature corresponding to the total condenser pressure as measured by the vacuum gauge. Finally, condensate coming into contact with the air will have its dissolved oxygen content increased.

For efficient condensing a number of design factors must be considered. A large heat exchange surface offering as little resistance as possible to heat flow should be provided to facilitate the rapid exchange of heat between the exhaust steam and the cooling water. This can be done by providing a large number of small diameter, thin walled tubes. Condensers are normally arranged to give cross flow heat exchange conditions which, provided a change of state is taking place, such as in the condensation process, gives a similar efficiency to that of a counter flow heat exchanger.

As the exhaust steam gives up its latent heat its temperature remains constant but the temperature of the cooling water rises as it flows through the tubes. It is therefore necessary to maintain sufficient temperature difference along the tube length for optimum heat transfer. To achieve this the cooling water should circulate through the tubes at as high a velocity as is practicable. As well as giving a good rate of heat transfer and keeping condenser size down, it also helps to prevent silting and fouling in the tubes. The maximum velocity is limited by the ability of the tube material to withstand erosion. When the coolant is circulated by a pump, its velocity can be varied to compensate for changes of sea temperature.

Care should be taken in the design of the circulating system so that it does not offer undue resistance to flow, and that water is evently distributed across the tube plate so that all the tubes are cooled efficiently.

On the steam side, no undue resistance should be set up to the exhaust steam, and its velocity across the tube bank should remain constant, as if this is allowed to vary less efficient heat exchange occurs, and there will be a greater tendency for any air to form pockets in the tube nest. Well designed condensers achieve this by steadily reducing the flow area available to the steam as it travels through the condenser.

The construction and materials used should be such that the condenser can resist corrosion and erosion on both steam and water sides, and does not leak, thus avoiding contamination of the feed water.

Methods used to support the condenser must allow for expansion, and any relative movement between the low pressure turbine and the ships structure.

AUXILIARY CONDENSERS

These consist of surface type heat exchangers basically arranged to a cross flow configuration. The sea water coolant passes through a large number of small bore tubes, while the exhaust steam supplied to the shell circulates around the outside of these tubes. The resulting condensate collects at the bottom of the shell, then passing to the hot well.

In general terms, thermal efficiency is not a primary consideration indeed many of the other factors previously mentioned may be neglected in the design of auxiliary condensers, provided they can cope with their most extreme operating conditions. These would occur at maximum steam load, while the coolant circulating pump is operating at full power with the highest expected sea temperature, this latter reducing the available heat flow between the exhaust steam and the cooling water.

The general arrangement of an auxiliary condenser is shown in Fig. 11. The tubes, usually of aluminium brass, are attached to the vertical tube plates of rolled admiralty brass, by ferrules and suitable packing rings. A number of steel support plates are fitted along the length of the tubes, while steel bar stays are fitted between the two tube plates. Suitable cap nuts are used to protect these stays from the sea water where they pass through the tube plate. The water boxes may be of cast iron or fabricated mild steel, in many cases lined internally with a bonded rubber or plastic coating to protect them against sea water corrosion. The water boxes and tube plates are connected to the fabricated steel shell by special collared studs (see Fig. 12) which permit the water boxes to be removed without disturbing the tube plates.

In the design shown, the cooling water only makes a single pass through the condenser, but in larger versions a two pass arrangement may be used if desired. This is done by fitting a division plate in one water box, while the coolant is supplied and removed from the other.

Air vents are fitted to the highest points of the water boxes and to division plates where fitted. A drain valve is also fitted so that once the appropriate valves have been closed the sea water side can be completely drained.

Baffles should be fitted where the various drains enter the condenser shell to

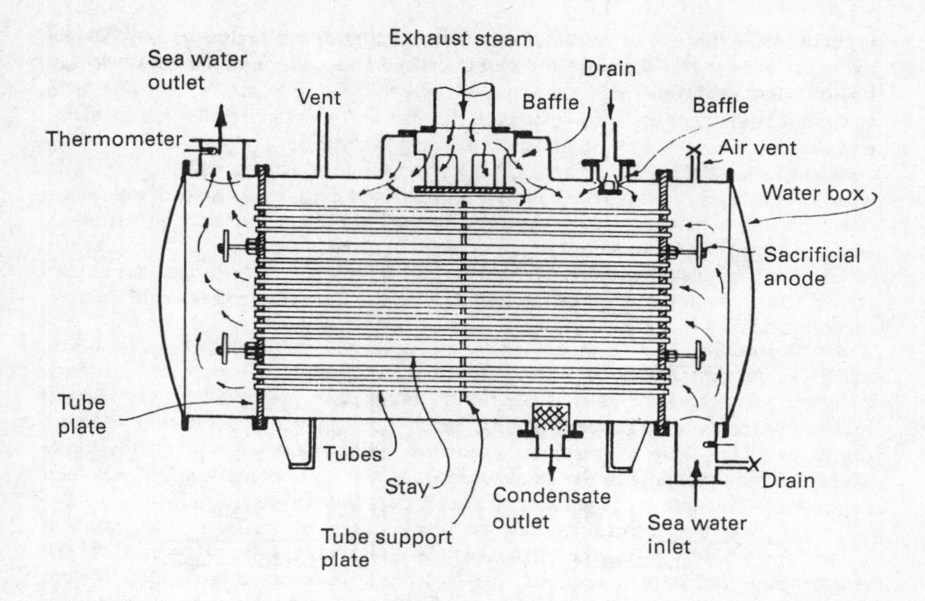

Fig. 11 Single Pass Auxiliary Condenser

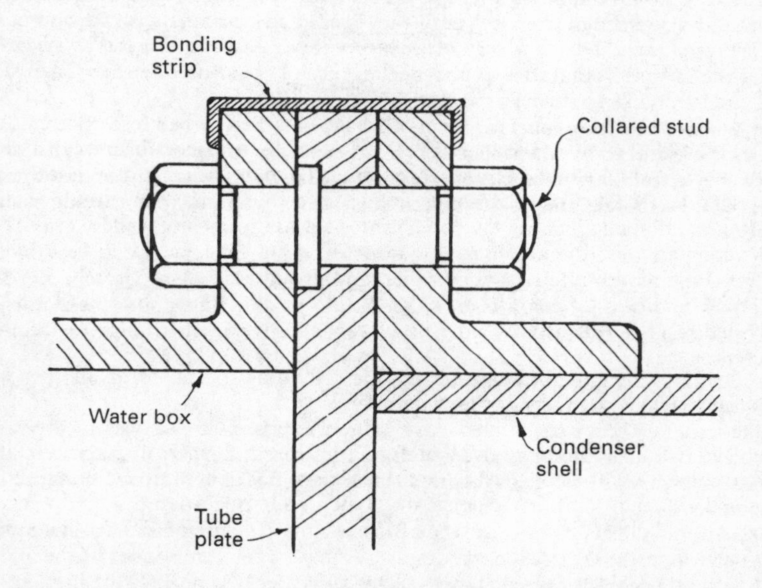

Fig. 12 Tube Plate Stud

prevent steam and water returns directly impinging upon the tubes. The type of deflector shown in Fig. 13 is recommended as this can easily be removed for inspection or renewal.

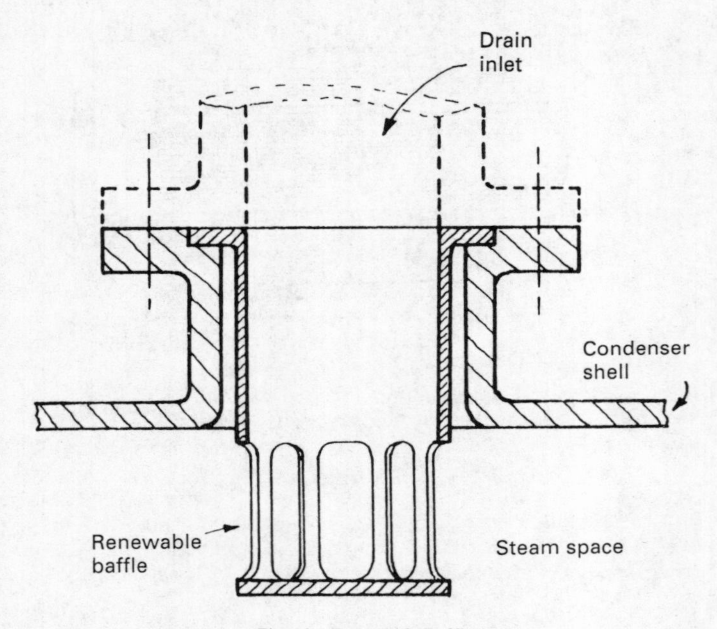

Fig. 13 Renewable Baffle

The type of condenser shown in Fig. 11 is suitable for use with simple auxiliary plant where only a slight vacuum is required. In many cases it may not even be necessary to fit any form of air or condensate pump, the very small vacuum existing in the shell allowing the condensate to drain to the hot well by gravity.

When, however, the auxiliary condenser is designed to work in conjunction with turbine driven alternators or other similar units, it will operate under high vacuum conditions. Separate take off points for air and condensate must be provided, and it will usually now be of regenerative form similar to the main condenser.

REGENERATIVE CONDENSERS

When used for the main engines or for large electric generating sets, the condenser must be carefully designed to enable a high thermal efficiency to be obtained for the plant. A regenerative type condenser attempts to do this by using a number of special features to meet the design factors for maximum efficiency previously considered.

A cross section through a typical regenerative condenser is shown in Fig. 14. The regenerative effect is obtained by means of a passage through the tube bank

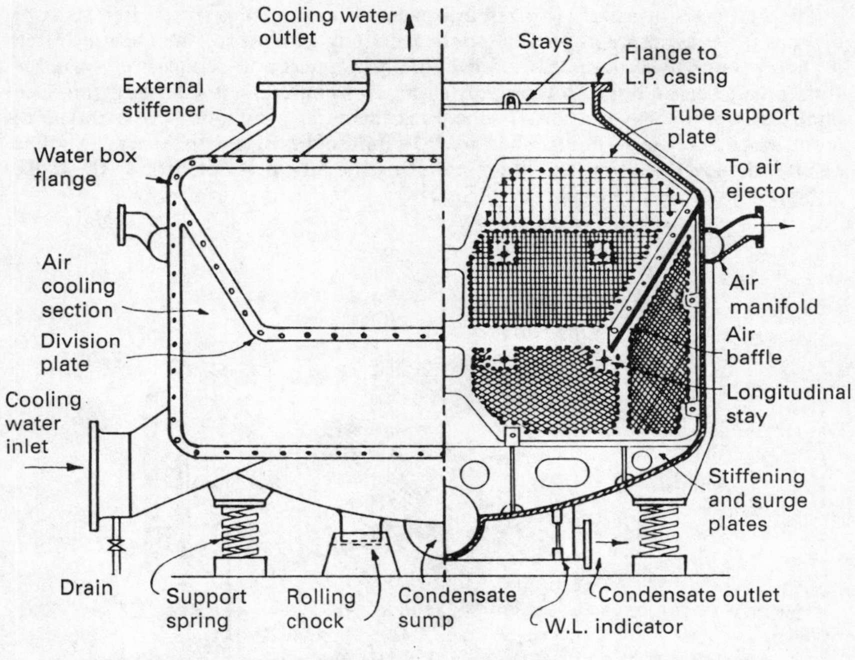

Fig. 14 Regenerative Condenser, Underslung Type

which allows a proportion of the exhaust steam to reach the lower parts of the condenser. Droplets of condensate dripping from the tubes then mix with this relatively hot steam, which gives up its latent heat to reheat these water droplets by direct contact, thus reducing the amount of undercooling to within 1°C of the exhaust temperature. This reheating also provides a deaerating effect; as when the water droplets are raised to their corresponding saturation temperature, dissolved gases in the water will be driven out. Suitable baffles then direct these gases to an air cooling section, from whence they can be drawn off by the air ejector. This arrangement can reduce the normal dissolved oxygen content of the condensate leaving the condenser at less than 0.02 ml/litre.

A manifold is fitted to ensure these released gases are drawn off evenly along the full length of the air baffle. It should be noted that the air take off points are placed a few tube rows below the top of the air cooling space, so allowing an air pocket to form which insulates the outgoing gases from the relatively hot baffle. This prevents them from being re-expanded as they leave the condenser.

The various return drains should be introduced into the upper part of the condenser to ensure the hot water entering also undergoes this deaerating procedure. Where these drains enter the condenser, deflectors should be fitted to prevent direct impingement of water onto the tubes causing erosion.

The first rows of tubes are often arranged with a coarse vertical pitch so as to offer as little resistance as possible to the incoming exhaust steam. The tube pitch is then progressively reduced and eventually staggered to reduce the available flow area. This is done to maintain a steady steam velocity through the tube bank. However this variation of tube pitch increases production costs and so in many cases a constant tube pitch is used. In some cylindrical condensers a similar result is achieved to some extent by mounting the tube nest eccentric to the shell, although this is mainly to obtain the regenerative effect.

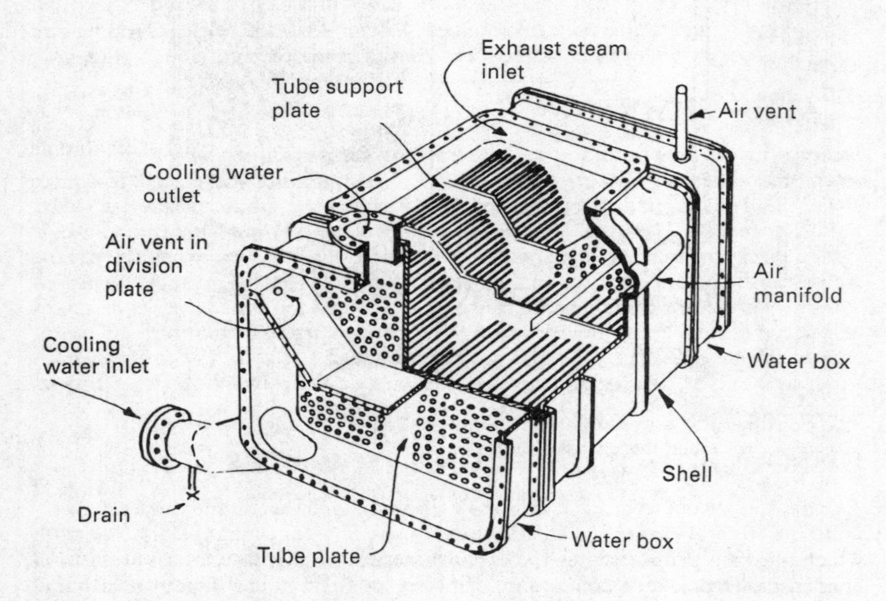

Fig. 15 Two Pass Regenerative Condenser

The condenser illustrated in Fig. 15 allows the cooling water to make two passes, the division plate following the line of the air baffle, so that water on its first pass is circulated through the close pitched tubes in the air cooling section below the baffle, thus obtaining maximum possible cooling effect. There is however an increasing use of single pass condensers, this being due to the greater use of scoop circulation where it is important to offer as little resistance as possible to the flow of cooling water through the condenser.

The cooling water should flow through the numerous thin walled tubes with as high a velocity as practicable for the operating conditions. This normally entails water speeds in the order of 2 to 4 m/s. The ability of the tube material to withstand erosion finally limits the speed. Cupro nickel tubes offer the best resistance and can operate at water speeds of over 10 m/s as compared to about 5 m/s for aluminium brass. However cupro nickel is both more expensive and offers greater resistance to heat transfer than the aluminium brass, and so is not generally fitted in merchant ship condensers unless exceptionally high water speeds are to be used or long periods of operation in highly polluted water are expected.

The water speed through the condenser tubes should not be allowed to fall below 1 m/s otherwise undue build up of deposits may occur leading to poor rates of heat transfer, and to possible corrosion of the tubes.

Basic constructional details and materials are similar to those stated for the auxiliary condenser. Differences which do exist include the possible use of cupro nickel tube plates and (or) tubes. In a main condenser the latter are usually expanded into the tube plate at the inlet end while the outlet end may also be expanded, or alternatively secured by means of a ferrule using alternate fibre and metallic packing rings as a seal. When this form of packing is used the metallic rings help to reduce the risk of corrosion due to galvanic action by providing an electrical connection between the tubes and tube plates. See Fig. 16.

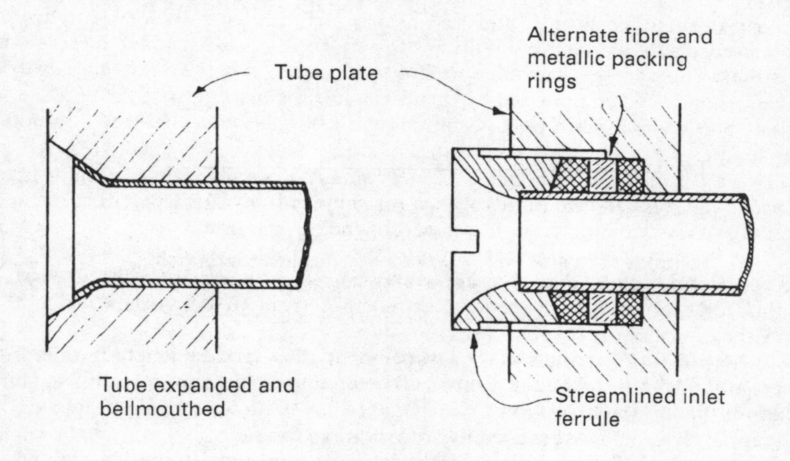

Fig. 16 Condenser Tube Attachment

Using ferrules at one or both ends, allows for expansion of the tubes through the tube plates. The latter can thus be supported by longitudinal steel stays. Cap nuts are used to protect these from the sea water. Provision is also made for these stays to be tightened up, or removed, without disturbing the tube plates. One method of doing this is shown in Fig. 17.

This form of stay cannot be used when the tubes are expanded at both ends, as provision must now be made for the tube plates to move relative to each other and provision for expansion made in the shell. The tube plates now rely for support entirely upon the expanded tubes or alternatively by stays fitted in the water boxes, with suitable protection from the sea water.

A number of mild steel tube support plates are placed along the length of the tubes. The holes in these plates have to be oversize to allow the tubes to be inserted, and in some cases severe wear on the outside of the tubes can occur in way of these support plates due to vibration.

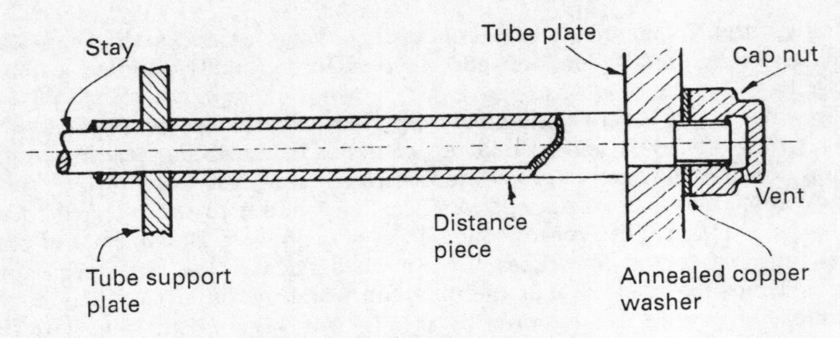

Fig. 17 Tube Plate Stay

There are two main methods of positioning the condenser relative to the low pressure turbine. The underslung arrangement as shown in Fig. 14 provides the exhaust steam from both ahead and astern turbines with a short path to the condenser, although the fact that it has to turn through 90° tends to increase the possibility of windage. This term refers to a process whereby exhaust steam is entrained and then compressed by the astern turbine blading so causing a loss of power, and coupled with poor vacuum conditions this can also lead to unduly high temperatures in the astern turbine.

The length of the turbine set is kept as small as possible but its height is increased, and this can be a disadvantage for engine aft layouts where the fine lines limit the space available in the lower part of the engine room.

In the version shown support springs are used; these take about two-thirds of the condenser weight and give good allowance for expansion. An alternative method of support is to suspend the condenser from strong girders which also carry the low pressure turbine casing.

Fig. 18 illustrates a single plane arrangement. This reduces height but tends to increase the length. It makes it more difficult to provide a path for the exhaust steam, although the axial flow arrangement shown allows a straight path to the condenser. This helps to reduce power loss due to windage.

Care must be taken with single plane layouts to prevent an unduly high water level in the condenser from spilling over into the turbine and causing damage. It is thus usual to fit both low vacuum and high water level alarms to this type of condenser.

Some form of sea water circulating pump must be fitted; even when using a scoop circulating system which relies upon the passage of the hull through the sea to induce a flow of water through the condenser, a circulating pump is still required for low speed conditions. However the size and power of this pump will be much smaller than one required to handle circulation at full load.

Either centrifugal, or axial flow pumps are suitable, in most cases being electrically driven.

An overboard discharge together with high and low injection valves are fitted. These are often of butterfly type in order to offer as little resistance as possible to the flow of circulating water. The low injection is normally used, this being placed low enough not to be exposed during heavy rolling. It should be noted that both injection and discharge valves should be opened before starting the circulating pump, especially when this is of axial flow type as these, unlike

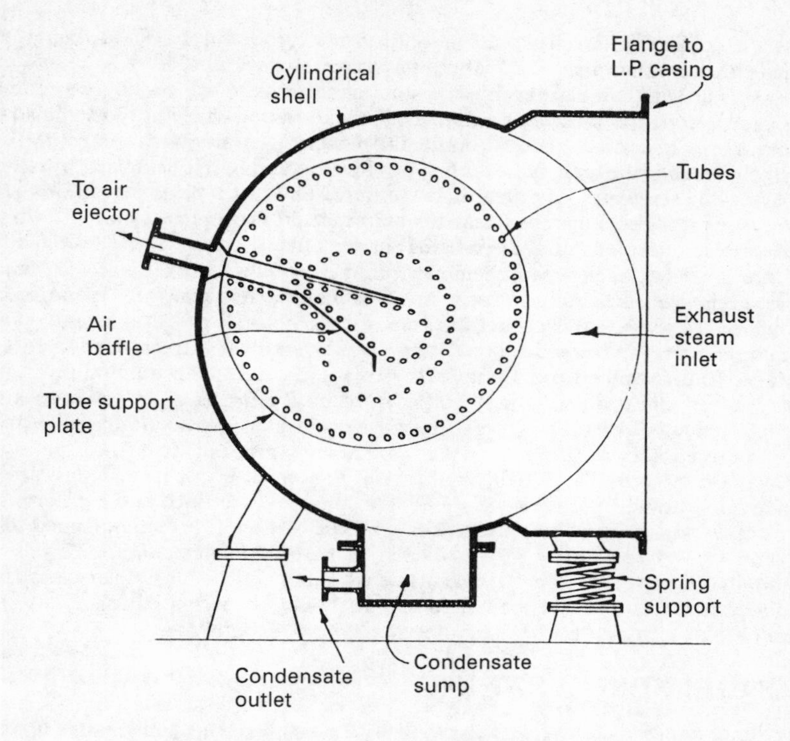

Fig. 18 Regenerative Condenser, Single Plane Type

normal centrifugal pumps, produce high pressure and starting torque at zero flow condition.

An emergency bilge suction is also fitted, so that in the event of serious flooding of the engine room, this emergency valve can be slowly opened while shutting in the main injection. This procedure continuing until the circulating pump is drawing partly or completely from the engine room bilge, and the water level is held steady just below the level of the bottom plates until the inflow of water can be staunched.

PROTECTION OF CONDENSERS

Avoid having circulating water speeds too low as silting can occur, or too high as this can cause tube erosion.

Cathodic protection should be provided for tubes and tube plates by fitting soft iron or mild steel sacrificial anodes in the water boxes when these are of non-ferrous material, or protected by a bonded rubber coating. These anodes provide a protective film inside the tubes. The effect can be increased by using an impressed current. In this case larger anodes, sometimes of a different material, are fitted.

Another method of providing a protective film on the inside of the tubes is to dose the cooling water with a 10% solution of ferrous sulphate, this being injected either continuously or for about an hour each day.

Marine growth can be prevented by adding chlorine to the cooling water and this can be done by injecting a suitable chemical, taking care not to use poisons whose use is restricted for environmental reasons. An alternative is to use an electrolytic chlorine generator. This produces sodium hypochlorite by electrolsying the sea water, giving a dosage rate of about 0.5 parts per million. In some cases it is recommended that the chlorination process is turned off while the injection of ferrous sulphate is being carried out.

Care must be taken when chemical treatment of the cooling water is being carried out to ensure that none of the treated water enters sea water evaporators being used to provide fresh water for domestic purposes.

Tube erosion can be reduced by fitting the inlet ends with streamlined ferrules, or if expanded by bell mouthing them. In some cases, with aluminium brass tubes, additional protection may be provided by the fitting of plastic inserts at the inlet ends of the tubes. These inserts, held in place by a suitable adhesive, should project about 150 mm into the tube and be tapered off to a fine edge.

Stagnant sea water is particularly harmful to condenser tubes, and shut down condensers should have their sea water side completely drained and kept dry. In some cases where ferrous sulphate dosing has been used, it is recommended that the condenser be refilled with fresh water and stored in this condition so as to maintain the protective film formed inside the tubes. When for shorter periods it is not practicable to drain the condenser, it should be circulated for about one hour each day, preferably by water dosed with ferrous sulphate.

CONDENSER CLEANING

Shut and secure the shipside valves, then drain and open up the water boxes, checking that they are empty and valves tight shut before the last nuts are removed. If the condenser is spring mounted, check before draining or flooding is carried out, for any special chocking arrangements which may be necessary to avoid straining the springs or the joint to the low pressure turbine casing.

Note the general condition inside the condenser before cleaning commences, then place protective covers in the bottoms of the water boxes to avoid damage to the bonded lining. The tubes can then be cleaned by means of water jets or by plastic balls blown through with compressed air. Only use brushes or flexible canes as a last resort, as these tend to damage any protective film formed on the tube surface. When plastic inserts are fitted always work from the inlet end. If brushes are used, push right through and withdraw from the outlet end to avoid damage to the inserts. When cleaning has been completed, dry off the tube plates and if necessary test for leakage. Plug or repair any leaks detected and retest. Do not disturb stay cap nuts unless there is positive evidence of them leaking. In this case it may be necessary to remove and anneal the copper washers fitted between the nut and the tube plate. Clean any sacrificial anodes by wire brushing and renew them as required, usually after about 50% of their original mass has been lost.

The protective cover sheets can now be removed and the condition of the water boxes checked. Make sure the division plate, if fitted, is intact and repair any damage to the bonded coating with a suitable patching compound. Make sure air vents and drains are clear.

Inspect the steam side for any deposits formed on the outside of the tubes. If bad, these may be removed by boiling out with a suitable solvent. Clean the condenser sump of any deposits and inspect and clean any filters fitted. Examine air baffles and any drain deflectors, repairing or renewing as necessary. Look for signs of erosion or vibration damage to the tube surfaces. Check for possible air leakage in way of joints, valves, or their glands.

Inspect the emergency bilge suction and make sure it opens and closes easily. If a strum box is fitted make sure it is clear. See that the circulating pump and pipelines are in good order.

When work is complete, see that all tools and other items have been removed and box up. Just before warming through of the main engines commences, open the shipside valves and refill the condenser ready for use. Close the drain valves and remove any chocks, which may have been fitted, at the appropriate time.

CONDENSER LEAKAGE

Sea water contamination of the feed system will be indicated in a number of ways, such as an increase in the salinometer reading, and by various changes in the results of the routine tests carried out on samples of feed and boiler water. Another sign in some cases is a rise of the water level in the reserve feed tank not in keeping with normal operation.

Once these indications have been checked out and the source of contamination has been isolated to the condenser, small leaks can often be cured by the addition of sawdust to the circulating water. This can be done either by means of a special hopper and injection line or, if not fitted, via the auxiliary circulating pump suction valve chest.

Meanwhile prepare to shut down and rectify the leakage. Check over the procedure for testing and repair of the condenser, and make sure all the necessary equipment is ready to hand. Then if the chloride level in the boiler water cannot be controlled and the sawdust ineffective, inform the bridge and shut down. With high pressure boilers it is imperative that any leakage is rectified as soon as possible and only delayed where the loss of steerage way would directly hazard the ship. Lower pressure boilers can tolerate larger amounts of contamination and, where it is possible to operate the boiler within the prescribed limits, it may be possible to reach port before shutting down.

After cooling down, the shipside valves for the circulating system can be shut and secured, the condenser drained and, following a similar procedure to that previously given, the water boxes opened up. If necessary fit seals to the turbine glands and carry out a suitable test procedure. Any points of leakage should be marked when located. A temporary repair to any leaking tubes can be made by driving a wooden or brass plug into the tube ends, or by screwing in cap nuts in place of the ferrules where these are fitted. The condenser can then be boxed up and refilled, any chocks or turbine gland seals that have been fitted removed, and the engines returned to service with notification to the bridge that the vessel can proceed on her way.

TEST PROCEDURES

A number of these are available, of which the following are the most widely used. The actual test chosen depends upon the equipment and time available and, to some extent, the type of condenser. For example there will normally be no need to carry out an elaborate test procedure on, say, a small auxiliary condenser where slight leakage can be tolerated. It can be tested by simply opening the water boxes and drying the tube plates then, by filling the steam space with water, any relatively serious leaks will immediately become evident. If it is desired to trace smaller points of leakage, then one of the tests can be carried out as for a main condenser.

ULTRA-SONIC TEST

Condenser should be drained and any necessary access doors removed. Electrical tone generators are then placed in the steam space of the condenser, care being taken to position some in the air cooling section. The sound produced by these passes through leaks to be detected by an ultra-sonic probe moved over the tube plate. It should be possible to actually insert the end of the probe into the tube ends so as to locate points of leakage more accurately. The probe can be tuned so as to pick up only the note produced by the tone generators, ignoring other machinery sounds.

A variation on this basic procedure is to subject the steam space to vacuum conditions or, if not convenient, to a slight air pressure and then use the probe to detect the sound of air passing through any leaks.

FLUORESCEIN TEST

Drain the condenser and remove any necessary access doors. The tube plates and tubes should then be cleaned in the normal manner. If the condenser is spring mounted, chocks must be fitted to support it and the steam space flooded up to the level of the top tubes with water containing a small amount of fluorescein. This will turn the water green. The tube plates and tubes are then illuminated by means of an ultra-violet light. Any points of leakage will then be indicated by a fluorescent glow. Care should be taken when adding the fluorescein to ensure that none is split over the condenser where it could lead to false indications of leakage. After testing is complete the water should be dumped. If left in the system, one problem to arise would be the distortion of the results of any colour comparitor tests carried out on samples of boiler water.

VACUUM TEST

Drain the condenser and remove any necessary access doors. Seals are fitted to the turbine glands as required and a vacuum provided in the steam space. This is done by partly filling the condenser with condensate, and using the extraction pump to circulate this through the air ejector then back to the condenser via the recirculating line. Steam is then supplied to the ejector nozzles to provide a suitable vacuum for the test. Leaks can now be detected by spreading thin sheets of polythene over the tube plates, Depressions in these plastic sheets indicate points of leakage. Alternatively an ultra-sonic probe can be used as previously described.

If during this procedure overheating should occur, add more condensate from the reserve tank or stop testing and allow to cool down. This precaution is especially important if plastic tube inserts are fitted.

CONDENSATE PUMPS

In simple auxiliary condensers working virtually at atmospheric pressure, the condensate will drain to the hot well by gravity. When vacuum conditions are involved some form of pump must be fitted to remove both the condensate and, as previously noted, air from the condenser.

The main problem here arises from the very low suction head available. The use of this term can be misleading as it gives the impression that the pump is responsible for actually sucking the liquid into the pump chamber whereas, in fact, it is the pressure acting upon the free surface of the liquid which pushes it through the pipe line to the pump. The difference in the pressure produced at the pump suction, to that acting upon the free surface, provides the energy to supply the liquid to the pump. The pressure in the condenser is low and therefore very little energy is available for this purpose. In order to keep friction and other losses to a minimum the suction line from the condenser to the pump should be as short and direct as possible, and have a bore at least equal to that of the suction branch on the pump.

As the condensate will approach its saturation temperature, any reduction in pressure caused by frictional resistance to flow in the pipe line, or by too great a variation in the pump speed, can lead to some of the liquid passing into its vapour state. This formation of vapour inside the pump chamber can, in reciprocating type pumps, cause heavy knocking at the end of the stroke or water hammer in the pipe lines. In centrifugal pumps it can cause the pump to gas up and lose its suction and also encourages the onset of cavitation which may possibly cause erosion damage to the impeller.

Therefore, while direct acting or reciprocating type pumps, often referred to as air pumps, can be used for moderate vacuum conditions, their piston speeds must be limited to prevent undue changes of velocity in the liquid flow to the pump. For these very low suction heads, displacement pumps are fitted with large plate or kinghorn valves, which enable large flow areas to be obtained. These valves have a small lift, and no springs are fitted as there is not enough head available to lift the plate against a spring.

For higher vacuum conditions the pump will need assistance to remove air and vapour from the condenser, and this can be provided by some form of air ejector. In this case a separate take off point for the air and vapour will be required, while some form of centrifugal extraction pump will now be fitted to deal with the condensate. These pumps running at constant speed do not produce the velocity fluctuations experienced in the suction line with reciprocating type pumps.

CHAPTER 3

Extraction Pumps

Extraction pumps are fitted to remove the condensate from the condenser sump and then to provide sufficient discharge head to overcome the various pumping losses in the system, yet still giving the necessary pressure to deliver the feed water to the feed pump inlet or to the deaerator.

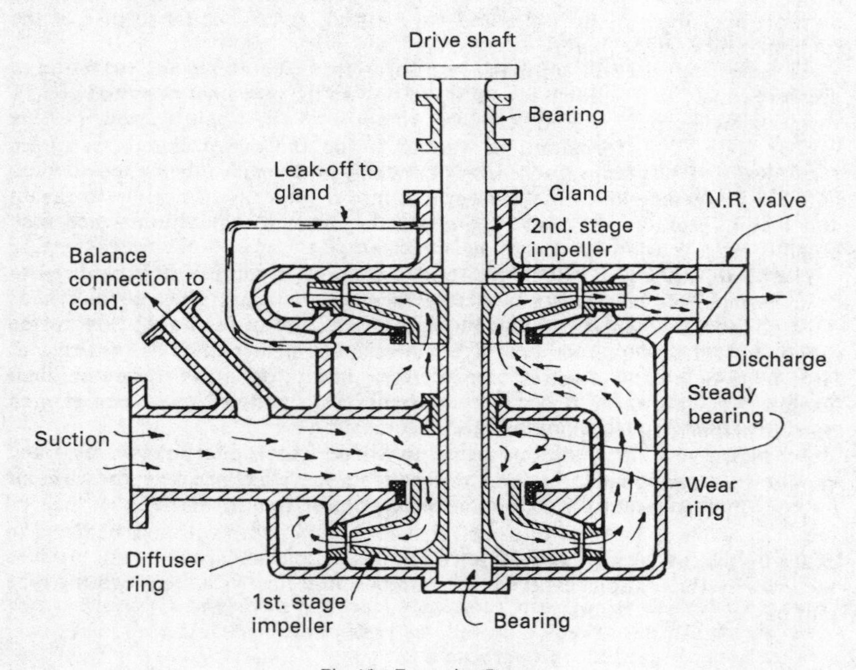

Fig. 19 Extraction Pump

As shown in Fig. 19, extraction pumps usually consist of two stage centrifugal pumps driven by a constant speed electric motor. The first stage impeller raises the condensate from condenser pressure to just above atmospheric, while the second stage provides the required discharge pressure. As can be seen in Fig. 19, the gland is placed on the discharge side of the pump to prevent any air from leaking in while the pump is running. This gland may be sealed with soft packing, but most modern pumps use some form of mechanical seal.

The stainless steel pump shaft is supported by three bearings, the internal one being water lubricated. The two aluminium bronze impellers are mounted facing each other to give some measure of hydraulic balance.

Due to the vacuum conditions in the condenser these pumps must be designed to operate with very small suction heads. The problem being increased by the limited distance between the condenser bottom and the tank top. Thus the extraction pump can only be placed a small distance below the level of the condenser sump although a slight improvement may be obtained, in some cases, by mounting the pump in a recess below the level of the tank top.

There are two basic designs to cope with these problems:

CONSTANT HEAD TYPE

These are capable of operating with a small but constant suction head. This is achieved by maintaining a constant water level in the condenser sump by means of a float controller.

This type of extraction pump runs at about 1200 rpm and can give a maximum discharge pressure in the order of 600 kN/m^2. As they are very sensitive to changes in the suction head they must be mounted as near to the central, vertical, athwartship plane of the condenser as possible, to minimize the effect of heeling.

SELF REGULATING TYPE

These are used in conjunction with the so called dry bottomed condenser, where the water level in the sump is allowed to vary. There is thus no need to fit a float controller. When manoeuvring, the water level may surge over the lower tubes or fall until the sump is empty. This arrangement therefore requires the fitting of a pump carefully designed so that its output varies to suit the available suction head.

This is done by fitting a two stage pump whose first stage is designed to allow cavitation under controlled conditions. When the suction head falls to a certain minimum value (usually about 0.5 m) the pump fully cavitates, vapour bubbles forming in the first stage impeller, and the discharge rate falls to zero. As the level in the condenser rises the increasing suction head reduces the amount of cavitation, and the discharge rate of the pump increases at a steady rate, without gulping, to reach a maximum output for a head of about 1 m. No further increase in the discharge rate of the pump will then take place, even if the level in the condenser surges up over this level for a short time. For normal running conditions the pump will maintain a water level in the condenser which leaves about half the bottom vapour passage clear.

These cavitating pumps are similar in general layout to the previously described constant head type but run at higher speeds, in the order of 1800 rpm, and have a modified first stage. They can give a maximum discharge pressure of about 850 kN/m^2.

When the vapour bubbles which form during cavitation implode onto pump surfaces, damage normally results but at the very low pressures involved in the first stage of a self regulating pump, so little energy is released that damage is minimal. It can be further reduced by designing the first stage impeller to be supercavitating, which results in the bubbles collapsing clear of the impeller surfaces.

Diffuser rings are also fitted. This is because in centrifugal pumps running at high speeds, low rates of flow can lead to the pressure distribution around the impeller being so uneven as to set up radial forces that can cause rubbing between the impeller and its wear rings. The fitting of a diffuser ring prevents this.

Dry bottomed condensers are normally used when a deaerator is included in the system. A buffer tank fitted below the deaerator now acts as the surge tank, thus releasing the condenser sump from this duty. The sump can thus be made shallower so reducing the overall height of the condenser. This is also one of the reasons why the arrangement is always used in conjuction with single plane layouts. As these are very sensitive to high water levels in the condenser, the sump level is often so low that the extraction pump continually cavitates to some extent, and for this reason a supercavitating type pump is often fitted.

OPERATION

Slight leakage should always be allowed through the gland to lubricate the packing. With mechanical seals this leakage may be so slight as not to be noticeable.

Some form of water seal must be fitted to the glands of the extraction pump suction valves, and is sometimes also fitted to the gland of the stand by pump. These seals should be frequently checked to make sure they are supplied with water and effectively preventing the ingress of air to the system.

When the pumps are opened up, the impeller wear ring clearance should be inspected. This should be in the order of 0.025 mm/25 mm diameter, with a minimum value 0.125 mm. The rings should be replaced when the clearance reaches about four times its original value. The impellers and casing should also be examined for signs of erosion. However the cavitation occurring at the very low pressures experienced in the first stage of these pumps should not normally cause damage of this type for the reasons previously considered. It should be noted that this is not the case in pumps working at higher pressures.

The non-return valve fitted at the pump discharge should be checked both for condition and its free movement.

CHAPTER 4

Air Ejectors

Air must not be allowed to accumulate in a condenser for the reasons previously considered. Thus for condensers working at high vacuum conditions, an efficient means of air removal must be fitted. In the majority of cases this consists of a steam jet operated air ejector, although in a few cases the use of a liquid sealed rotary vacuum pump, driven by an electric motor, is preferred.

STEAM JET AIR EJECTORS

These convert the heat energy of high pressure steam into kinetic energy by expanding it through a convergent divergent nozzle, so designed that a small proportion of the expansion takes place after the steam has left the nozzle. This causes pressure waves to be set up, the resulting turbulence creating regions of low pressure into which particles of air and vapour from the condenser can flow. Here they are entrained by the high velocity steam jet and carried into the diffuser, where the kinetic energy is converted to pressure energy, so that the air and vapour is discharged from the diffuser at an increased pressure.

The mixture then passes through tubes in the cooling section, the feed water on the other side of these tubes acting as the coolant. Much of the vapour, together with the operating steam, is thus condensed and returned to the system via a suitable steam trap. The air and any remaining vapour continue on to the next stage. This and any subsequent stages which may be fitted perform in a similar manner, the air finally being discharged to the atmosphere. Each stage of the ejector can be made smaller than the previous one as the volume of the air and remaining vapour decreases with the rise of pressure.

A non-return valve must be fitted to retain the vacuum in the condenser by preventing air from flowing back through the ejector in the event of the supply of operating steam to the nozzles being shut off or otherwise interrupted.

Other fittings include a steam strainer in the steam supply line to the nozzles to prevent chokage. Air vents fitted to suitable positions on the feed water side permit the removal of air, which would otherwise form pockets so reducing the rate of heat transfer. Pressure relief valves must be fitted to the ejector where necessary.

Many variations in the design of steam jet air ejectors exist, the two types considered here however illustrate the basic principles.

THREE STAGE TYPE

This consists of three stages mounted in series and with internal diffusers. The general arrangement is shown in figure 20.

The air and vapour enter the first stage suction at condenser pressure, about 3.5 kN/m^2, or expressed as a vacuum, equivalent to 760 mm Hg. They are then

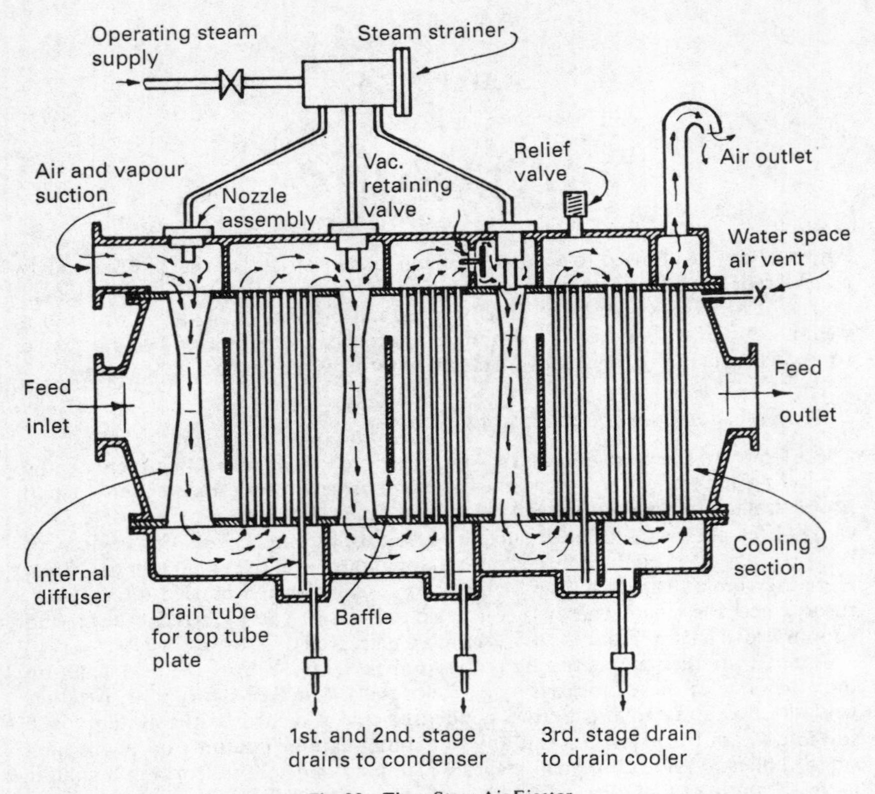

Fig. 20 Three Stage Air Ejector

pumped to the second stage at a pressure of about 28 kN/m^2 or 560 mm of vacuum and pass through the non-return vacuum retaining valve to enter the third stage. Here the air and any remaining vapour are discharged at a sufficient pressure to make three final passes through the cooling section, after which the air is released to the atmosphere.

Condensate from the first and second stages is returned to the condenser via a suitable steam trap, often a barometric one. The third stage drains to the drain cooler. A tube in the cooling section of each stage is extended downwards into a drain pocket, thus providing drainage for the top tube plate.

The operating steam to all the nozzles is controlled by a single valve. If the third stage cooling section or the air outlet pipe should become obstructed, the operating steam to the third stage nozzle could not pass back to the previous stages because of the vacuum retaining valve. A relief valve must therefore be fitted to this stage to prevent the possibility of overpressure.

On the feed water side a number of internal baffles are fitted to improve the flow pattern through the cooling sections and so obtain more efficient heat exchange. Air vents are fitted to enable any air to be removed from beneath the top tube plate.

Monel metal nozzles are fitted to mild steel holders while the internal gun metal diffusers, and the aluminium brass tubes, are expanded into the top and bottom mild steel tube plates. The whole being contained within a cast iron body. The vacuum retaining valve consists of a stainless steel plate valve on a monel metal seat.

TWO STAGE TYPE

This version has the diffusers mounted externally and although basically consisting of two stages actually has two of each mounted in parallel, so arranged that either first stage can be used in conjunction with either, or both, second stages. Under normal operating conditions only one first and one second stage would be in use. To enable this to be done, valves are fitted so that the redundant stages can be isolated. This type thus gives greater operational flexibility and standby capability than the previous design.

The general arrangement is shown in Fig. 21. The air and vapour enter the first stage at condenser pressure and are then raised to a pressure of about 35 kN/m^2 in order to enter the cooling section which provides a common suction to either second stage. Here the pressure is raised sufficiently to discharge the air into the atmosphere through the vacuum retaining valve.

The first stage drain is returned to the condenser via a suitable steam trap while the second stage drains to the drain cooler. A relief valve must be fitted to the second stage to prevent overpressure in the event of the vacuum retaining valve jamming in the closed position, while a closed isolating valve prevents the nozzle steam from escaping back to the condenser. If a valve is fitted between the condenser and the first stage nozzle, then an additional relief valve must be fitted to the first stage.

Similar materials to those of three stage are used, except that the casing is now of fabricated mild steel. The U-tubes fitted in the cooling section help to reduce the leakage problems often associated with straight tubes fitted between two tube plates.

PACKAGE UNITS

In many cases the air ejector is packaged into a single unit with the gland steam condenser, and in some cases a drain cooler. The basic operating princples however remain unchanged.

OPERATION

Eroded or coated nozzles cause loss of efficiency. They should therefore be inspected whenever possible and cleaned carefully, using only a grease solvent and a soft cloth to avoid damage. The nozzle should be renewed if it shows any signs of damage or erosion. The relative position of the nozzle exit to the diffuser, is critical and even too thick a joint in the cover can lead to loss of efficiency. Ensure that the correct nozzles are fitted for each stage. When this is being done the steam strainer should also be examined and cleaned.

It should be noted that supplying more steam to the ejector nozzles will not necessarily increase the condenser vacuum. Thus, after full away, when the job has settled down, slowly close in the ejector steam supply until the vacuum

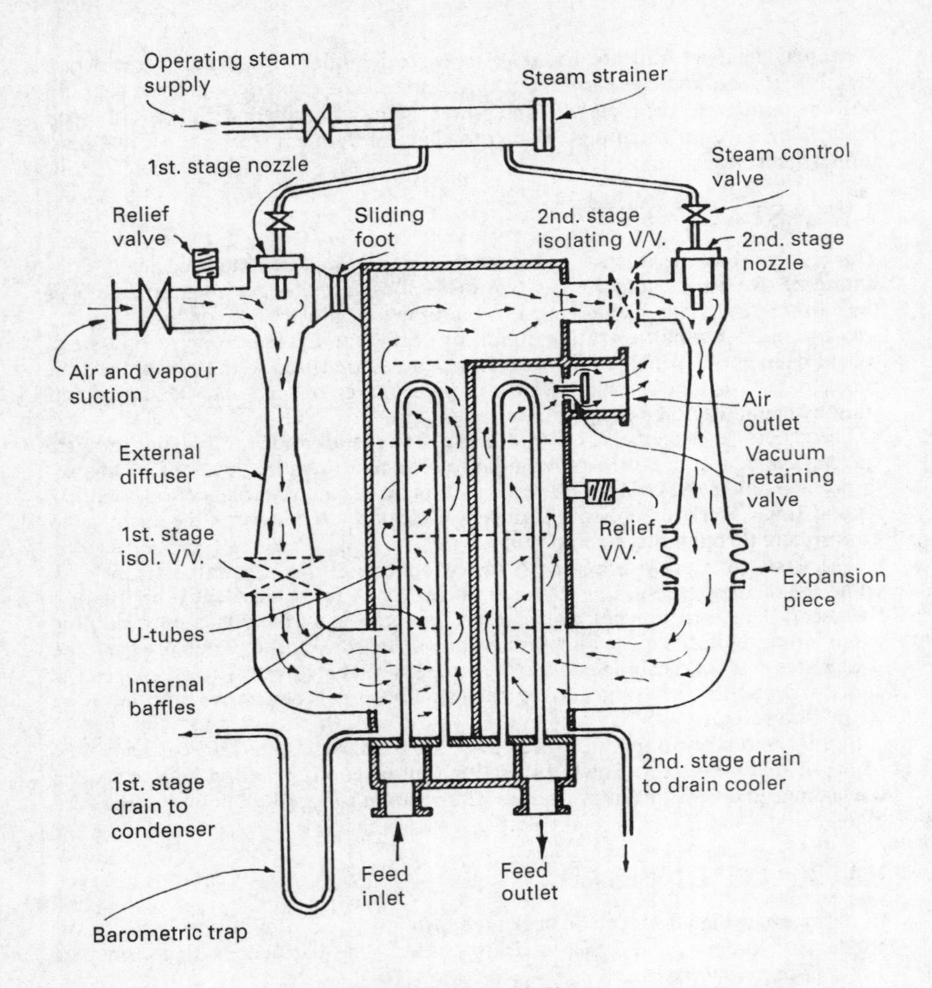

Fig. 21 Two Stage Air Ejector

begins to fall. Then reopen the steam supply valve just sufficient to restore the vacuum. Any more operating steam would be wasted. If more steam than normal is required the reason for this should be investigated. Among other things it can indicate air leakage into the condenser, or loss of efficiency at the ejector nozzles.

When shutting down, check the vacuum retaining valve for flatness and free movement, by placing a sheet of paper or thin card, over the air outlet as the operating steam is turned off. If there is any tendency for the paper to be drawn in, it indicates that the non-return valve is leaking.

Tube leakage can be checked by circulating feed water through the cooling section of the shut down air ejector by means of the extraction pump, running with the recirculating valve open to prevent the pump from overheating. A constant flow of water from any of the ejector condensate drain outlets will then be an indication of leakage.

If air leakage through a cover joint is suspected, it can be checked out on a shut down ejector by means of a halogen test. The procedure for this is to insert a gas detector into the air outlet from the ejector. Then a halogenated gas, of which freon is an example, is sprayed onto the outside of the ejector in way of possible points of leakage. Any indication of gas by the detector will be a sign of leakage through the joint.

Make sure the condensate side of the ejector is vented when starting up, and occasionally when operating, to remove any air which would otherwise reduce the rate of heat transfer in the cooling section.

RECIRCULATING VALVE

To prevent steam jet operated air ejectors from overheating, a supply of coolant is required at all loads, including manoeuvring conditions. A recirculating valve must therefore be fitted at some point in the feed line after the air ejector which, by allowing a proportion of the feed water to be returned to the condenser, ensures sufficient continuous circulation through the air ejector and the gland steam condenser if fitted, to cool them even when the boiler feed regulators have stopped the flow of feed into the boiler.

For manual operation this recirculating valve is opened at standby and closed at full away. When an automatically operated valve is used, the control action can be governed by either a flow sensor or a thermostat mounted at a suitable point in the feed line.

ROTARY AIR PUMPS

Electrically driven vacuum pumps may be used in place of steam jet ejectors to remove air. Water sealed rotary air pumps are particularly successful in handling the saturated air and water vapour mixture from the condenser. Compared to steam jet ejectors they give faster air removal, and so smaller air cooling sections can be used. Additionally there is no need to fit a high pressure steam supply line and control valve. It is also claimed that they make more efficient use of the energy from the fuel by allowing steam at high pressure and temperature to be expanded efficiently through a turbine driven alternator so producing electricity for the pump motor, than they would by expanding the steam through a nozzle and reclaiming the heat by the feed water cooling the ejector. The elimination of a steam jet ejector from the system enables some other type of low temperature heat exchanger, such as a distiller, to be put in its place. A disadvantage is that the electric motor and rotary pump form a more complex unit likely to require more maintenance.

The layout of a water ring type vacuum pump is shown in Fig. 22. The rotor is mounted eccentrically within a circular chamber partly filled with water. The rotor contains a number of vanes curved in the direction of rotation, and shrouded at the sides to form a series of pockets. As the rotor revolves it causes the water to form a ring concentric with the casing, thus forming a series of spaces around the central part of the rotor. Due to the eccentricity between the water ring and the rotor, these spaces first increase in size. This causes air and vapour to be drawn in through the suction port, then after moving past the top dead centre position these spaces progressively reduce in size so expelling the air and some vapour through the discharge port into the atmosphere. Thus each pocket in the rotor draws in and expells a charge of air and vapour during each revolution. Much of the vapour is condensed by the water ring during this process, excess water being removed from the casing as necessary.

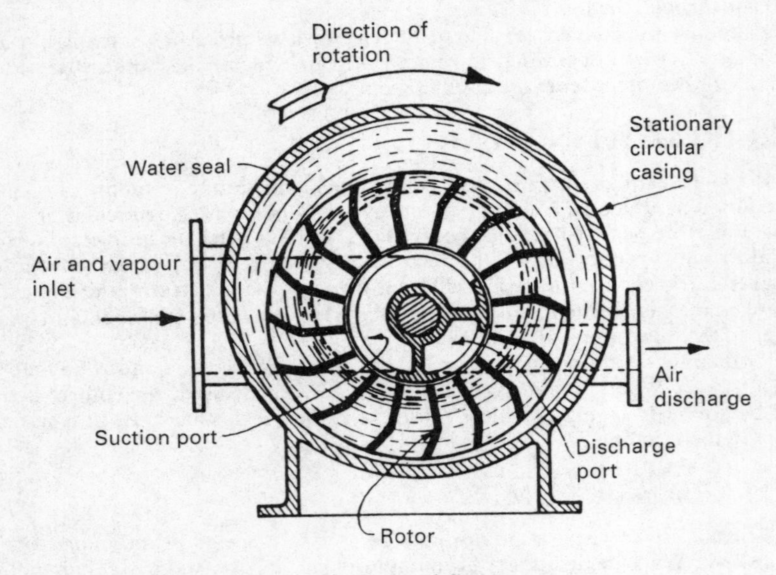

Fig. 22 Rotary Air Pump

CHAPTER 5

Low Temperature Heat Exchangers

The condenser, extraction pump and air ejector must form a basic part of any main feed system. However it is usual, in order to increase the thermal efficiency of the plant, to also include various low temperature heat exchangers in the feed system. These reclaim the heat in exhaust steam, hot water draining from various sources, and return it to the feed water. They also help to prevent the loss of distilled water from the system. The most important of these units are considered in this chapter.

GLAND STEAM CONDENSERS

These receive vapour from the turbine glands vapour collector, condense the steam and return the condensate to the feed system via a barometric trap. Any air will be discharged into the atmosphere. These units thus not only reclaim heat, but conserve feed water and reduce engine room humidity since if they are not fitted the vapour is discharged directly into the engine room atmosphere.

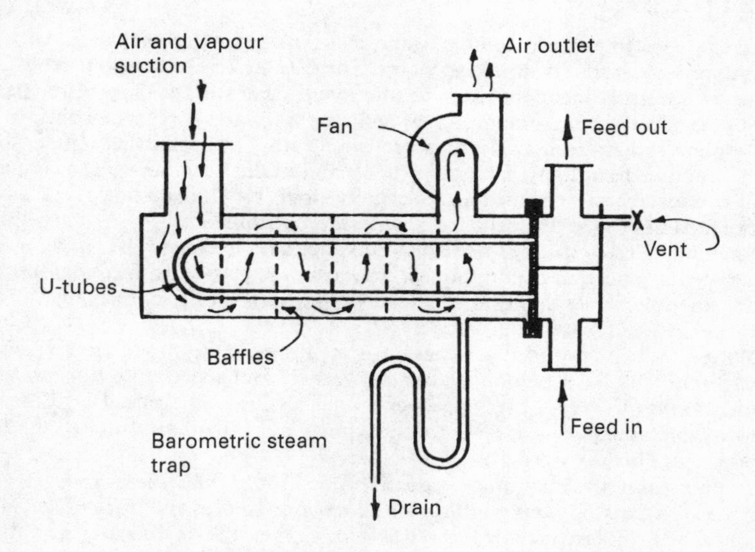

Fig. 23 Gland Steam Condenser

Fig. 23 shows the general arrangement of a gland steam condenser fitted with an electrically driven fan. A flow of vapour is induced through the cooling chamber by maintaining it under slight vacuum conditions, either by means of a fan as shown, or by a steam jet ejector. In the latter case a second cooling chamber is required to condense the operating steam. Another arrangement sometimes used is where an electrically driven fan is fitted to provide a small positive air pressure in the turbine gland vapour hoods. Thus the vapour is forced up to, and through the gland steam condenser. This now works at a slight positive pressure and no fan or ejector need be fitted to it.

The gland steam condenser is constructed of fabricated mild steel. A number of internal baffles are fitted in the cooling chamber so that the vapour has to make a number of passes over the aluminium brass U-tubes. These are expanded into the mild steel tube plate. As previously noted this unit is often incorporated in the same casing as the air ejector, and in some cases it also acts as a vent condenser for the deaerator, the vent outlet from this being led into the cooling chamber.

OPERATION

The tubes should be checked for leakage in the same way as those of the air ejector. Vents and drains should be proved clear and if necessary cleared. The heat exchange surfaces may also require occasional cleaning.

In the event of the exhaust fan failing the unit will usually continue to function reasonably well.

DISTILLERS

These are used to provide potable water, this term referring to evaporated water used to provide domestic drinking water. The distiller receives vapour from a sea water evaporator and condenses it, the resulting distillate then being passed through a filter section containing animal charcoal and limestone chips. When the vapour comes from a low pressure evaporator, the operating temperatures will be too low to kill any bacteria present, and in this case the water leaving the distiller must be sterilized. This is usually done by chlorinating the distillate before it passes to the domestic fresh water storage tanks.

When included in the feed system the distiller uses the feed water as a coolant, usually being placed in circuit immediately after the gland steam condenser. This arrangement however makes the distiller output very sensitive to changes in feed temperature and flow conditions.

When the distiller forms a separate unit it should be mounted high in the engine room or have a loop placed between it and the evaporator, to avoid contamination by water carrying over to the distiller. The raised position also gives the advantage that fresh water leaving the distiller can drain to the fresh water storage tanks by gravity.

In other cases the distilling condenser is combined with the evaporator, and consists of a nest of aluminium brass tubes mounted in the upper part of the vapour shell. When included in the feed system the feed water circulating through these tubes acts as the coolant.

OPERATION

The tubes should be tested to twice the pressure to which the relief valves must be set, or if none are fitted to twice the boiler pressure. The heat exchange surfaces should be checked at regular intervals and if necessary cleaned and any leaks rectified. The filter section should be cleaned out at least once every six months.

DRAIN COOLERS

These are responsible for cooling hot drains from various units to a temperature suitable for their return to the condenser or atmospheric feed tank, so reclaiming heat from them and reducing the heat that would be lost from the system if the hot drains were to be returned directly to the condenser.

Drain coolers consist of surface type heat exchangers, the feed water passing through tubes, while the hot drains circulate around the outside of the tubes. There are two basic types in general use:

FLOODED TYPE

These receive and cool hot water drains only. The general arrangement of a drain cooler of this type is shown in Fig. 24.

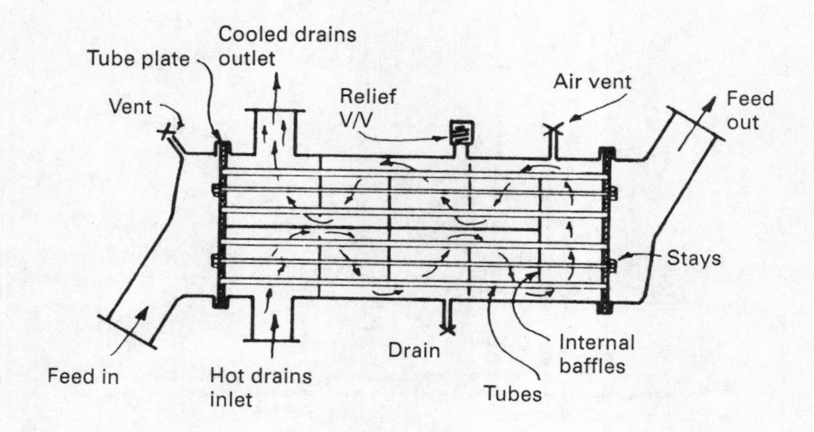

Fig. 24 Drain Cooler

The hot drains enter at the bottom of the shell and pass around numerous internal baffles. These are fitted to improve the heat transmission under the poor fluid to fluid heat exchange conditions existing in the cooler. The aluminium brass tubes may be U-shaped and expanded into a single mild steel tube plate, or straight tubes as shown, attached to the tube plates by ferrules and suitable packing rings. The tubes slide through these ferrules to allow for expansion. The shell is constructed of fabricated mild steel. Deflectors are fitted to prevent any water entering the shell with a high velocity, impinging onto the tubes.

VAPOUR TYPE

In addition to hot water drains this type also receives vapour from the vents of various units. The basic layout for a vapour type drain cooler is similar to that of the low pressure heater shown in Fig. 25. The vapour and hot water drains enter the upper portion of the drain cooler passing around the internal baffles until the vapour is condensed. The resulting condensate together with the drain water then undergoes further cooling in the lower portion of the shell, which is kept flooded by a standpipe in the outlet pipe. In many cases the shell is maintained under a slight vacuum condition by means of a continuous vent back to the main condenser. Relief valves must be fitted as necessary.

The materials used in the construction of vapour type drain coolers are as those used for the flooded type.

LOW PRESSURE HEATERS

These, in addition to the vapour and hot water returns, are also supplied with heating steam which may consist of exhaust steam from various units, or steam bled from the low pressure turbine casing. The general arrangement of a typical low pressure heater is shown in Fig. 25.

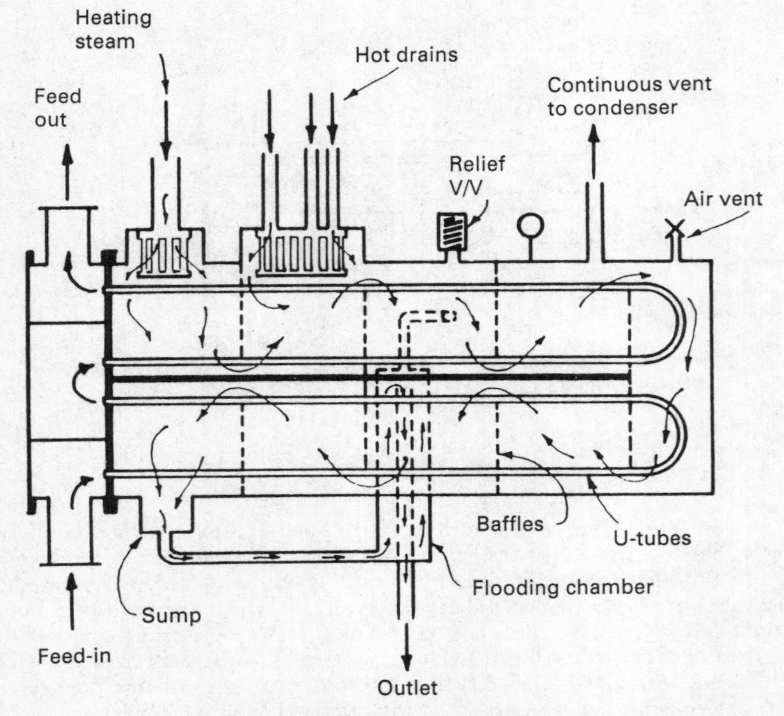

Fig. 25 Low Pressure Feed Heater

These heaters are fitted to raise the feed to a temperature suitable for entry into a deaerator, or in some cases to reduce the mass and thus the volume of steam flowing over the last rows of blades in the low pressure turbine. This helps to reduce the blade heights in addition to the gain in thermal efficiency. In this latter case no other heating steam or drains must be allowed to enter the shell, and no control valve is fitted between the turbine and the cooler.

If, during manoeuvring, vapour and hot drains are continually supplied to these units then in order to prevent them overheating, a flow of coolant must be provided during periods of intermittant feed flow. Thus it will be necessary to position the take-off point for the recirculating line after the drain cooler or low pressure heater concerned.

OPERATION

Check periodically for tube leakage. This can be done by a similar procedure to that used for the air ejector, that is by circulating feed water through the drain cooler with all the heating steam and water returns shut off. A constant flow of water from the shell drain will indicate leakage although this could of course be due to a leaking valve. Examine deflectors and make sure they are in good order, repairing or replacing them as required. Check for any signs of tube erosion, and make sure internal baffles are intact.

Keep steam and water sides clean, the former occasionally needing boiling out with a suitable solvent. If any vents are led into the shell, check for signs of corrosion. Any relief valves fitted must be tested as required.

CHAPTER 6

Regenerative Feed Heating

Regenerative feed heating is used to increase the thermal efficiency of the plant. Steam is bled off at some intermediate stage through the turbine or engine, and then used to heat the feed water in a contact, or a surface type feed heater. The general arrangement for this is shown in Fig. 26.

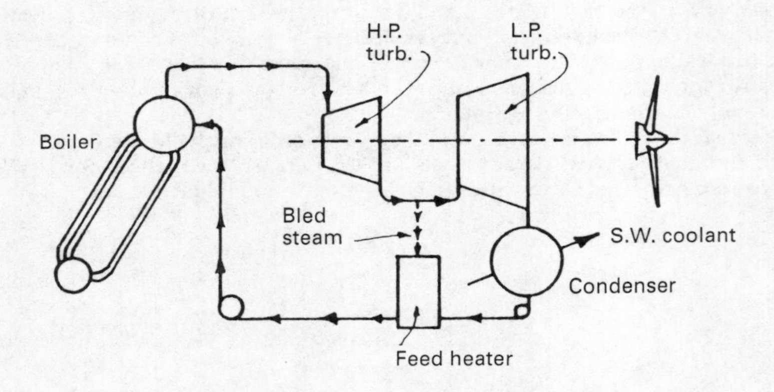

Fig. 26 Regenerative Feed Heating

If no feed heating is used, each unit mass of steam supplied to the main engine gives approximately three times as much energy to the cooling water flowing through the main condenser, as it gives to do useful work while being expanded through the turbine. Therefore, if some steam is bled off after having done some useful work, and then condensed in a feed heater, it now gives energy, mainly its latent heat, to the feed water serving as the coolant. Thus this energy is reclaimed in heating the feed water, whereas if given to the sea water cooling the condenser it would have been lost. Such saving of energy in the system is greater than the fraction of work lost by not fully expanding the bleed steam through the turbine.

DIRECT CONTACT FEED HEATERS

These heaters work at low pressure and are fitted on the suction side of the main feed pumps. Heating steam is obtained either by bleeding steam off from a suitable turbine stage, or as exhaust steam from various auxiliary units. This steam enters the mixing chamber where it comes into direct contact with a spray of feed water. The two mix together, the steam condensing so giving up its latent heat, which goes to raise the temperature of the feed water. This process continues until the water reaches the temperature of the heating steam. The resulting mixture of condensed heating steam and feed water leave the heater together, so no drain is required.

Due to the low pressures involved and the fact that the water leaving the heater will be close to the corresponding saturation temperature, this type of feed heater must be placed relatively high above the feed pump to avoid vapour forming at the pump suction. This can occur if sufficient pressure drop takes place at the pump inlet to lower the saturation temperature below that of the water temperature. Some of this water will then flash off as steam, which can then cause the pump to gas up and lose its suction.

The heater shell is constructed of fabricated mild steel, and the various valves and other fittings, of gun metal.

The water level in the storage tank placed below the mixing chamber is kept within the desired limits by a control valve fitted in the feed supply line to the heater. This type of heater is suitable for use with auxiliary feed systems used in conjunction with low pressure tank boilers.

DEAERATORS

With high pressure boilers it is essential to keep the feed water free of dissolved gases. To do this use can be made of the fact that water raised to its boiling point releases the bulk of any gases dissolved in it, which can then be removed.

In a closed feed system the regenerative condenser will remove the bulk of these dissolved gases, reducing the dissolved oxygen content of the condensate to less than 0.02 ml/litre. However for boiler pressures above 3000 kN/m^2 it is recommended (and for boiler pressures above 4200 kN/m^2 considered essential) that a deaerator also be fitted. These consist basically of a direct contact type feed heater fitted with suitable venting arrangements for the positive removal of any released gases.

In the deaerator shown in Fig. 27 the feed water passes through a number of tangential spray nozzles, mounted in parallel, which cause the water to enter the mixing chamber in the form of a series of thin conical films which soon break up into small droplets. This allows the heating steam to come into contact with a large surface area of water, thus giving good conditions for heat transfer and for the release of any non-condensable gases. The heated droplets then fall onto a series of trays which provide further deaeration as thin films of water cascade over the edges of the trays. In addition this arrangement also provides a water seal between the mixing chamber and the hot deaerated water in the storage tank.

To ensure the positive removal of the released gases a vent condenser is fitted to the vent outlet of the deaerator. This is cooled by the incoming feed water. A slight vacuum may be maintained in the cooling section by a steam jet ejector or by an electrically driven fan. Another arrangement sometimes used is to lead the vent to the cooling chamber of a combined vent and gland steam condenser, or into a feed tank as shown in Fig. 9. It is this fitting of a vent condenser which provides the basic difference between deaerators and the direct contact feed heater previously described.

An efficient deaerator in a closed feed system should be able to reduce the

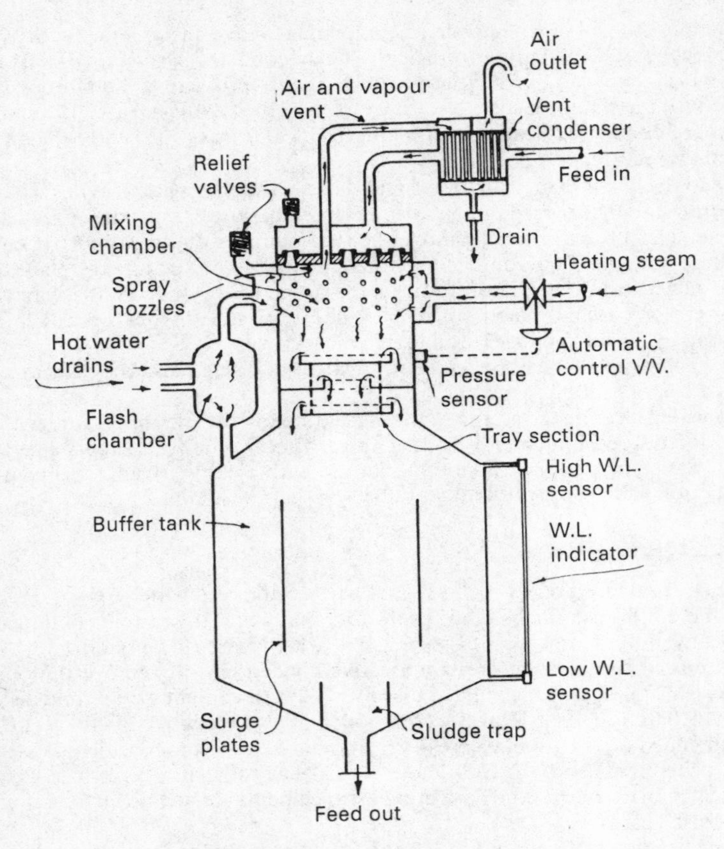

Fig. 27 Deaerator

dissolved oxygen content of the feed water to below 0.005 ml/litre, provided the temperature rise across the heater is at least 28°C and the oxygen content at the condenser outlet below 0.02 ml/litre.

The problem of gassing at the feed pump suction again arises as with contact feed heaters, and is usually overcome by mounting the deaerator high in the engine room so providing sufficient pressure head at the feed pump suction to prevent vapour forming. When it is not possible to place the deaerator high enough to obtain the necessary head, a boost pump can be fitted to the feed outlet from the deaerator to increase the discharge pressure. In some cases, especially when the deaerator is designed to work with the auxiliary system in port or where it is an addition to the original system, separate deaerator supply and extraction pumps are fitted. Another method of preventing the formation of vapour at the feed pump is to circulate the feed water on its way to the deaerator through tubes in the buffer tank, so cooling the water stored in it to a suitable temperature.

The heating steam can be obtained from various sources, the most common being bled steam from the turbine and exhaust steam from turbine driven pumps. An automatic control valve is normally fitted in the heating steam supply line in order to maintain a constant pressure in the mixing chamber. This in turn governs the temperature to which the feed water will be heated, i.e. the saturation temperature corresponding to the pressure. For example an absolute pressure of 170 kN/m^2 in the mixing chamber will give a corresponding feed temperature of 115°C. When a pressure of 340 kN/m^2 is used, the feed temperature is increased to 138°C. The higher temperature will increase the deaerating effect but will require heavier scantlings in the construction of the deaerator, so leading to greater weight and cost. In addition the problem of vapour at the feed pump suction will also be increased. On the other hand it should be noted that in order to prevent undue deposits forming on the gas side heating surfaces of the economiser, the feed water entering it should have a minimum temperature of 115°C. If the deaerator is capable of supplying this temperature then no further heaters need be fitted between it and the economiser as a basic requirement for the system. To enable this minimum required feed temperature to be maintained during manoeuvring, or at low load conditions, an additional source of heating steam should be available. This will be taken direct from the boiler via a number of pressure reducing valves, and used when the bled steam is shut off or reduced.

If sufficient high pressure, hot water returns are available, a flash chamber may be fitted. As the water enters this chamber it undergoes a drop in pressure which causes some of its mass to flash off into steam, which then adds to the heating steam entering the mixing chamber. The remaining water in the flash chamber will be at boiling point, thus dissolved gases will be released and the now deaerated water can go into the storage tank fitted directly below the deaerator. This tank serves as the buffer, or surge tank for the system, holding a quantity of hot deaerated water ready for immediate use. These tanks are relatively large often holding sufficient water for about ten minutes full power steaming.

The materials used in the construction of a deaerator are mainly fabricated mild steel with gun metal spray nozzles, and aluminium brass tubes in the vent condenser. Relief valves must be fitted to both steam and water sides.

In addition to being used in the closed feed systems of high pressure boilers, deaerators may also be included in open feed systems used in conjunction with auxiliary boilers working at pressures in the order of 2000 kN/m^2, especially when the boiler is of water tube type. The designs used in this latter case tend to be of more complicated form than the one shown in Fig. 27, which is suitable for a closed feed system.

OPERATION

As previously mentioned, the temperature rise across the deaerator should be maintained at a minimum of 28°C for the proper removal of oxygen. If in spite of this, and with no undue amounts of air entering the system, the oxygen content of the water in the buffer tank is too high, the following items should be checked: worn spray nozzles giving excessively large water droplets; damaged or displaced tray sections, or poor venting caused by dirty or choked tubes in the vent condenser. When the cooling section of the vent condenser is maintained at a slight vacuum condition, fan failure or worn nozzles in steam ejector types can lead to loss of efficiency. See that all steam traps and drains are clear. The buffer tank should occasionally be drained and the sludge trap at the bottom cleaned out.

SURFACE FEED HEATERS

In this type of heat exchanger the feed water passes through the tubes, while the heating steam enters the shell where it passes around the outside of the tubes until condensed, thus giving up its latent heat. The resulting condensate finally leaving the shell through a suitable drain connection.

As there is no direct contact between the heating steam and the feed water, large pressure differences can exist between them. This type of heater, therefore, can be fitted on the discharge side of the feed pump where the high pressure conditions enable the feed water to be raised to high temperatures without evaporation taking place. This is due to the fact that the saturation temperature increases to correspond to the high feed discharge pressure.

Surface heaters of the type shown in Fig. 28 may be used on their own, but in most cases are fitted in conjunction with deaerators and (or) economisers. If the latter are fitted the surface heaters may be necessary to raise the feed temperature to the required 115°C for the reasons previously mentioned. If this temperature has already been reached, the surface heaters can be used to increase the thermal efficiency of the plant by making greater use of steam bled off from the main turbine.

The temperature rise across each surface heater of the type being considered is usually in the order of 30°C. The cast steel header is subdivided (see Fig. 29) so that the feed water makes several passes through the heater; this gives higher water speeds thus enabling more efficient heat transfer to be obtained. The heat exchange elements consist of a number of U-tubes expanded into a single mild steel tube plate, so being free to expand. Earlier types of heater had straight tubes expanded into a fixed tube plate at one end, and attached to a floating box header, giving the necessary allowance for expansion, at the other. In some cases copper tubes are fitted but for high pressure conditions cupro-nickel or steel tubes are used. A monel metal, or spiral wound monel metal, and asbestos gasket are fitted in the joint between the tube plate and the header, which is subjected to full feed pressure.

The heater shell is of fabricated mild steel, with a number of internal baffles to give more efficient circulation of the heating steam. Deflector plates are also fitted to prevent direct impingement by the heating steam causing erosion of the tubes. The shell drain is normally fitted with a steam trap to ensure that all the heating steam is condensed before leaving the heater. At the same time excessive water levels in the shell must be avoided, as this would cause flooding of the lower part of the tube nest which would reduce the rate of heat transfer. A water level indicator fitted to the shell enables a correct level to be maintained in the shell. Relief valves must be fitted to both steam and water sides of the heater. Orifice plates may be fitted in the drain to maintain the required pressure in the shell.

Air vents are fitted to both steam and water sides. These should be opened when putting the heater into service and occasionally, during running, to remove air. If vapour from evaporators enters the shell to supply make up feed to the

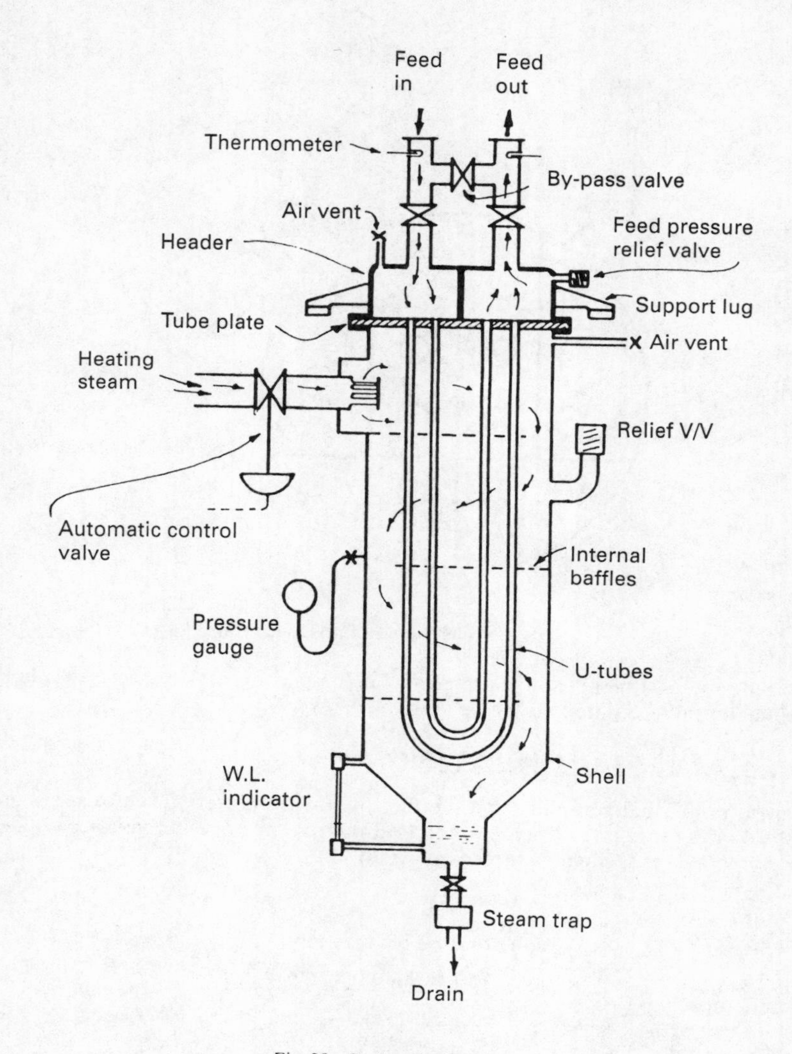

Fig. 28 Surface feed Heater

system, a continuous air vent should be fitted to the shell. This also applies if the deaerator vent is led into the shell, although this arrangement is not recommended as it can lead to excessive corrosion in the shell.

A by-pass valve is fitted on the feed water side to enable the heater to be isolated from the system. The bled steam valves fitted on the turbine casings must be of screw down non-return type to prevent water from entering the turbine in the event of a heater tube failure.

The heater must be tested, when new, to twice the boiler pressure plus twenty

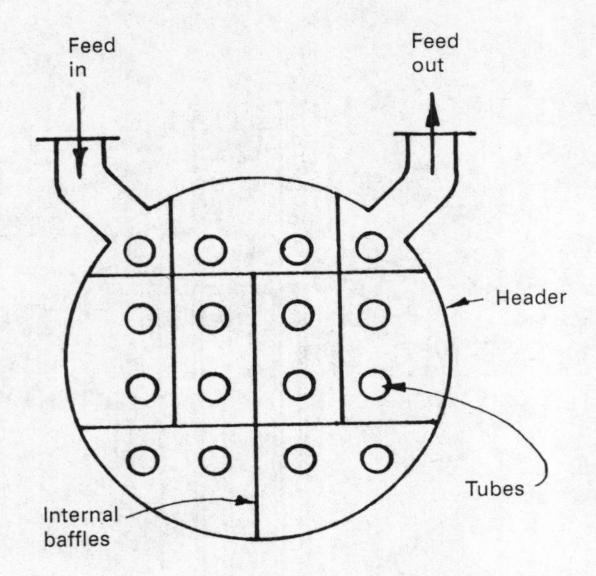

Fig. 29 Internal subdivision of Feed Heater Header

per cent and (or) be able to operate continuously at twenty five per cent above the maximum feed pump discharge pressure.

CASCADE FEED HEATING

This term is usually applied to a number of surface heaters fitted in series and placed in the feed circuit between the feed pumps and the economiser or boiler. The general arrangement is shown in Fig. 30.

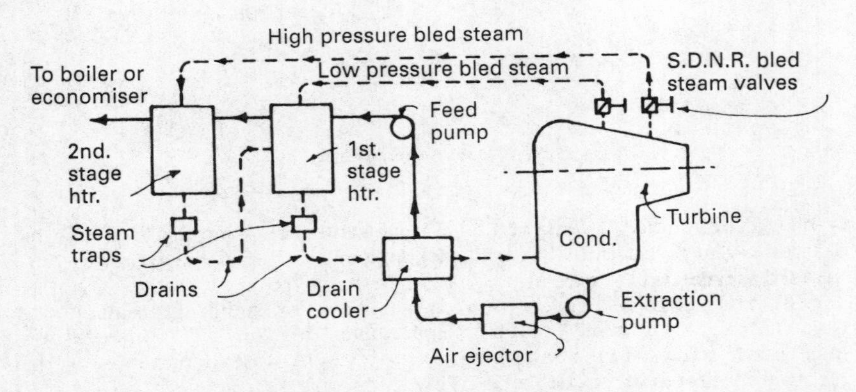

Fig. 30 Cascade Feed Heating

Each heater is supplied with bled steam from progressively higher pressure bleed points in the turbine. In addition the drain from each heater is led into the shell of the previous, lower pressure unit, so giving more energy to the feed water. This continues until the drain from the first stage heater is finally led to the main condenser via the drain cooler.

Theoretically each additional regenerative feed heater increases the thermal efficiency of the plant at the expense of some loss of power. However while each additional heater gives a progressively smaller increase in the overall thermal efficiency of the plant, the cost of each unit is basically the same. Thus the increase in capital cost sets a practical limit on the number of stages, and in marine installations cascade heating is usually limited to two or three stages. In a few cases cascade feed heating is used instead of economisers, the boiler efficiency being maintained at reasonable level by the use of gas air heaters.

OPERATION

With manual operation the bled steam to the feed heaters is put on at full away, the drains being adjusted to keep the water level in the gauges at half glass. At low loads or when manoeuvring, the bled steam is shut off to prevent undue temperature fluctuations in the heaters caused by intermittent steam and feed flows. This can result in the formation of steam in the heater tubes leading to water hammer which can cause leakage in way of expanded tubes or bolted fittings. In modern systems automatic control valves will be fitted to regulate this supply of heating steam to the shell.

Ammonia and oxygen in the heating steam can cause corrosion on the outside of copper or cupro nickel tubes, while carbon dioxide can lead to corrosion of the steel shell. Corrosion inside the tubes is rare, but may occur if the dissolved oxygen content of the feed water is high, while its pH value is low. Internal deposits are only occasionally found in the tubes; they are usually soft and can be removed by a suitable solvent.

To test for leaking tubes, shut off all heating steam to the shell and circulate feed water through the tubes. A constant flow of water from the drain of the water level indicator fitted to the heater shell, will then indicate tube leakage.

CHAPTER 7

Feed Pumps

Feed pumps must be capable of raising the feed water to a pressure high enough for it to enter the boiler as it is required.

There are two basic types of boiler feed pump in general use, namely centrifugal and displacement, the latter using buckets or rams. The pumps may be driven by direct acting steam engines, steam turbines, electric motors, or in a few cases from the main engines.

For auxiliary tank type boilers, relatively small amounts of water have to be handled. The pump may have to draw direct from double bottom tanks and must therefore be self-priming. The water demand by the boiler tends to be intermittent and displacement pumps, driven by steam or electricity, are best suited to meet these requirements.

Main or large auxiliary water tube boilers demand large quantities of steam under fairly steady flow conditions. The feed pump will have water supplied to it by an extraction pump, and so does not need to be self-priming. They must be able to raise the water to high pressures, and to work in conjunction with automatic feed regulators designed to maintain a constant water level in the boiler. To meet these demands, displacement pumps are occasionally used; e.g. a triple ram type pump driven by a constant speed electric motor. The pump output can be varied by altering the length of the ram stroke. This is done automatically in response to changes in the boiler water level. The pump itself controls the water supply, no regulating valve being fitted between it and the boiler (only the usual isolating and non-return valves).

However in the vast majority of cases high speed, multi-stage centrifugal pumps are preferred, as these can handle large quantities of water, while because of their high rate of throughput the size can be kept relatively small. They also provide steady flow conditions, so avoiding shock in valves and feed lines. These pumps are usually driven by steam turbines.

Some representative types of boiler feed pumps are now considered for main and auxiliary feed systems.

DIRECT ACTING FEED PUMPS

These steam driven displacement pumps are suitable for tank type auxiliary boilers. They give a high pumping efficiency over a wide range of throughput, and can deal with widely fluctuating feed demands. The double acting pumps are self-priming, and so can draw directly from double bottom tanks, but give an intermittent discharge action. While this is not a great disadvantage with small auxiliary boilers, it can easily lead to water hammer in valves and feed line; this limits the allowable piston speed resulting in low rates of throughput. Thus pumps of this type would be very large if used in conjunction with boilers having large evaporation rates.

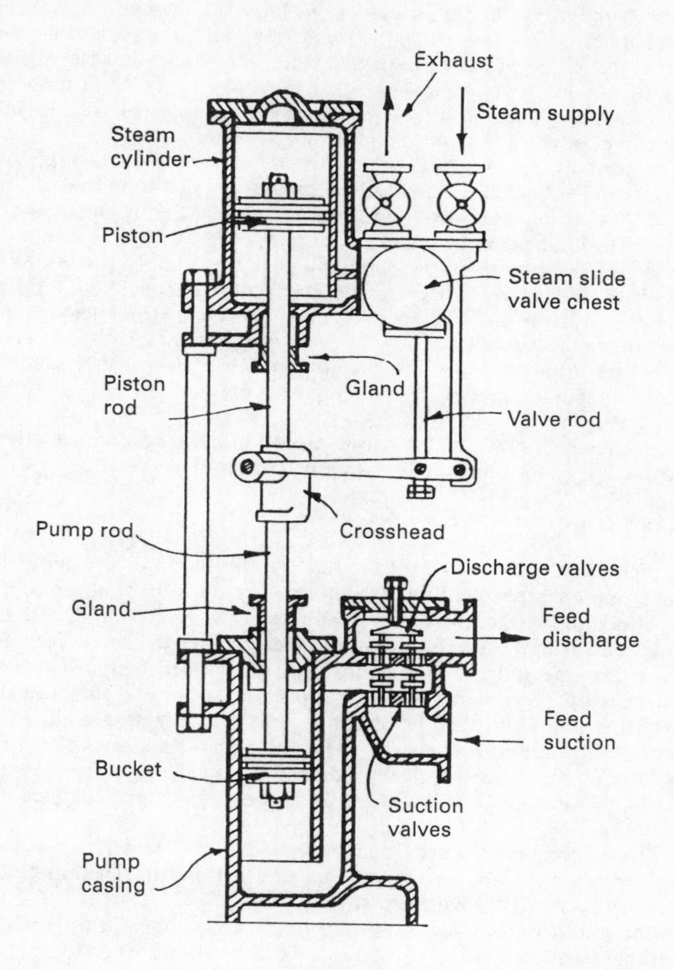

Fig. 31 Direct Acting Feed Pump

The pump shown in Fig. 31 has a single cast steel cylinder with a cast iron liner. The cast steel piston is attached to one end of the forged steel rod and the gun metal bucket to the other. The pump casing, also of gun metal, has a valve chest containing suction and discharge valves for both ends of the double acting pump. The valves consist of spring loaded, stainless steel plates. The piston rings are of cast iron, while the bucket rings are ebonite.

The piston and pump portions of the rod are screwed into a crosshead, the movement of which drives linkage used to actuate the cast iron steam slide valve. This usually supplies the double acting piston with steam for the full length of its stroke, although in some larger pumps of this type arrangements may be made to cut off steam admission to the cylinder part way along the stroke, allowing the steam already in the cylinder to partly expand. This both increases the thermal efficiency of the pump and allows the expansive working to give more gradual accelerations at the ends of the strokes, so reducing shock set up by rapid velocity changes of the water column in the feed line.

Steam is normally supplied at boiler pressure, and as the steam piston has a larger area than the bucket it allows a feed pressure higher than that of the boiler to be produced so that the water can enter as required. The steam end is constructed to withstand twice the boiler pressure so needs no relief valve, but one is normally fitted to the pump to prevent damage in the event of it working against a closed discharge valve. Theoretically the pump barrel should be able to withstand the pressure produced; in practice however, water leaking past the rings can allow the bucket to slowly move down its stroke. Thus the steam pressure acting upon the large piston area will force the small diameter rod further into the pump chamber, producing very high pressures if no relief valve were to be fitted.

The boiler feed water regulator can be used to operate a control valve in the steam supply line to the pump, so varying its speed as necessary.

OPERATION

To start the pump, open the water suction and discharge valves and if possible flood the pump chamber. Positive displacement pumps must never be operated against a closed discharge, therefore make sure boiler feed check is also open, or if a spring loaded by-pass valve is fitted, that it is set to the correct pressure. Open the steam line and cylinder drains together with the exhaust steam valve, then crack open the steam supply valve and allow steam end to warm through. The steam line drain can then be shut and that on the cylinder closed in. The steam supply valve can now be slowly opened to bring the pump up to the desired speed. If the pump is automatically controlled, open the steam supply valve slowly until the automatic valve takes over, then fully open the steam supply valve.

When direct acting pumps are fitted with a valve chest which allows the steam to be used expansively, the device must be shut off and the steam admitted for the full length of the stroke when starting up.

When the pump is running at normal load the cylinder drain valve can be closed although at low load conditions it would be advisable to leave it cracked open to keep the cylinder free of water.

Lubrication of the piston rod can lead to problems with oil contamination of the feed water due to the oil passing back to the auxiliary condenser with the exhaust steam from the pump.

Maintain steam and water valves in good condition and adjust steam and water glands at frequent intervals, renewing packing as necessary. Check that the rod in way of glands does not become unduly worn.

Excessive leakage past the bucket rings can cause the pump to operate erratically, and eventually to perform as a single acting pump, delivering a volume of water equal only to the volume of rod entering the pump chamber on the down stroke. The bucket rings are made of ebonite, and replacement rings should be soaked in boiling water for a few minutes to soften them before fitting. When new, the gap in these rings should be about 1 mm.

Too high a feed inlet temperature can lead to vapour forming on the suction stroke, so causing the pump to gas up. This can be cured by changing the suction to a reserve feed tank for a short time to cool the pump until the feed water temperature has been reduced.

RECIPROCATING FEED PUMPS

This type of pump, driven by steam or electricity, is often fitted for use in conjunction with small, low pressure boilers, steam generators, or evaporators. The pump shown in Fig. 32 is commonly used; it consists of a two throw reciprocating pump driven by a constant speed electric motor.

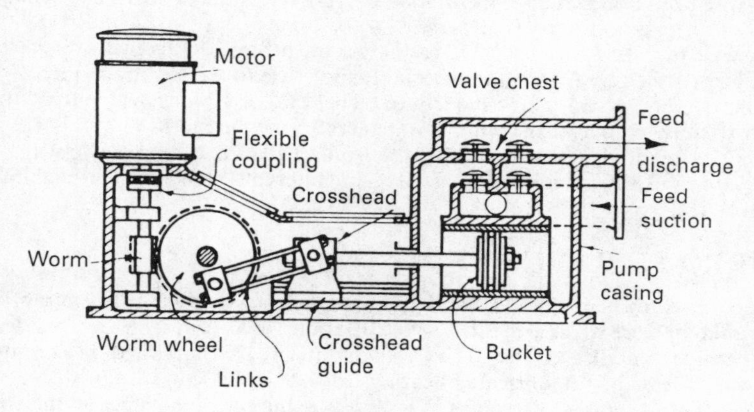

Fig. 32 Reciprocating Feed Pump

Compared to a direct acting pump, this type allows higher pump speeds to be used due to the fact that the crank mechanism gives more gradual speed changes over the ends of the stroke, producing smaller accelerations and so reducing shock in valves and feed lines. The two crank pins are set at 90° to each other which reduces fluctuations in the discharge pressure from the two double acting pumps. It should be noted that in steam driven pumps of this type the cranks must also be set at 90° to each other to ensure the pump can be started in any position.

The rotary drive of the electric motor is converted into reciprocating motion by means of a worm and wheel. This also enables relatively large speed reductions to be obtained without producing undue noise. The gearing is totally enclosed and runs in an oil bath. Rotating shafts, subject to torque only, run in ball bearings while bearings subjected to reciprocating forces are of plain white metal. The rest of the materials, and the general arrangement of each pump chamber is similar to the direct acting pump previously described, except in the smaller reciprocating pumps where rubber or neoprene rings are fitted to the buckets.

If the boiler water level is kept constant by means of a feed regulator the pump will run at constant speed, a spring loaded by-pass valve allowing excess water to circulate back to the pump suction when the regulator closes in. An alternative arrangement often used, where a fluctuating water level is permissible, is an on-off method of control. A float operated magnetic switch is fitted which, at a predetermined low water level, switches the pump on allowing it to run at constant speed supplying feed and so raising the water level until, at a predetermined high level, the pump is switched off.

OPERATION

To start the pump check that the lubricating oil level is correct and then open the pump suction and discharge valves. Power can then be switched onto the motor which, if automatically controlled, may not start immediately. It should therefore be put on manual control to make sure it is running correctly, being put back on automatic control if all is in order.

Instead of fitting a relief valve to the pump, protection against overpressure may be provided by fitting an overload trip or fuse to the electric motor. If this should operate, do not allow the trip to be held in, or a higher value fuse fitted, in an attempt to operate the pump without rectifying the fault.

Routine maintenance will consist of keeping valves and bucket rings in good condition and adjusting glands at frequent intervals renewing the packing as necessary.

TURBO-FEED PUMPS

The pump shown in Fig. 33 consists of a single stage centrifugal pump with hydraulic balance arrangements, driven at about 7000 rpm by a Curtis velocity compounded impulse turbine. The pump and turbine are mounted on a common shaft supported by oil lubricated bearings, one of which can be adjusted axially to maintain the correct clearance in way of the initial thrust rings on the shaft. These are fitted to deal with any small out of balance forces set up by disturbances in the hydraulic balance arrangements when starting or stopping.

Steam supplied at boiler conditions is expanded down to about 300 kN/m^2 through nozzles fitted in the top half of the turbine casing. The exhaust steam discharges into a back pressure line and can thus be used for feed heating. Exhaust at vacuum conditions would give a larger casing thus increasing the first cost of the pump. A pressure relief valve fitted to the turbine casing is set to lift about 100 kN/m^2 above the normal exhaust pressure.

As the turbine is of impulse type, no pressure drop takes place across the turbine blades and there is thus no need for fine mechanical clearances and the unit can be started up straight from a cold condition.

To keep length to a minimum, carbon glands are used to seal the turbine casing. The carbon segments in these glands are adjusted to very fine running clearances, and due to the different coefficients of expansion for carbon and steel, a mandrel is needed to enable the carbon segments to be set to their correct running clearance.

The pump casing shown is sealed by a single gland filled with soft packing. In the multi-stage versions required for higher discharge pressures, an additional end bearing will be fitted to support the longer pump shaft and two glands will now be necessary one at each end of the pump casing.

Fig. 34 shows the cross section through a typical pump gland. A small amount

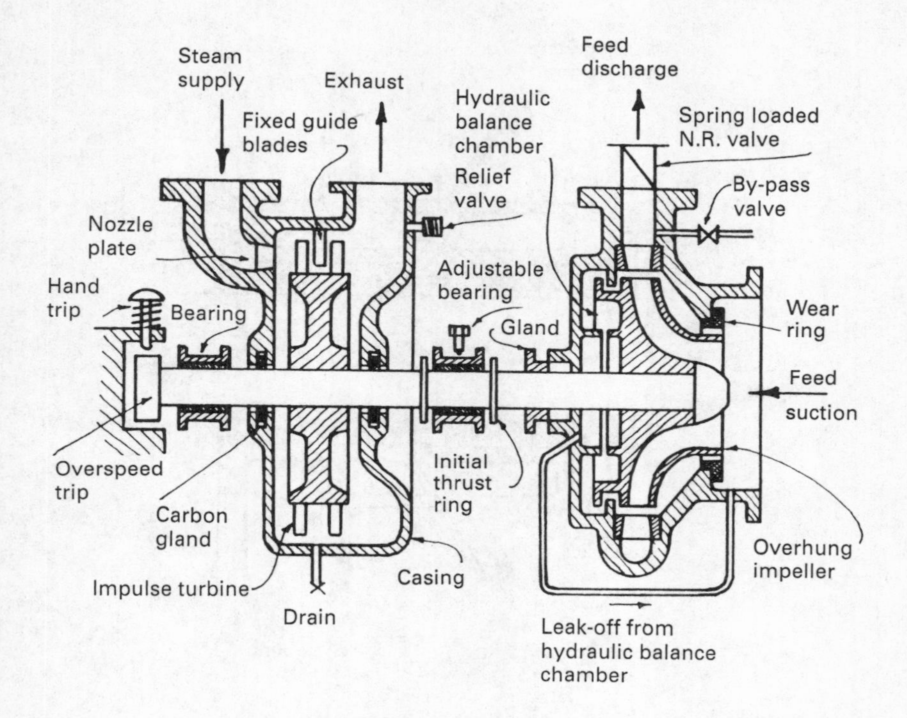

Fig. 33 Turbo-Feed Pump, Oil lubricated Bearings

of water must always be allowed to leak off through the gland when the pump is running, to cool and lubricate the packing. When high temperature feed water is being handled, this leak off may tend to flash off as steam leading to possible overheating of the gland. Cooling water supplied to the lantern ring from a suitable source, such as the extraction pump discharge, may be used to prevent this assisted if necessary by additional water supplied to a cooling jacket fitted around the gland.

To pack the gland the necessary valves must be closed to isolate the pump and the water drained off. The old packing can then be removed, extractors being used for the lantern ring. Check the condition of the shaft sleeve, and look for undue eccentricity of the shaft where it passes through the gland housing. New packing rings can then be inserted, each being pushed right home by means of suitable pieces of wood. After the correct number of turns have been fitted the lantern ring is inserted, care being taken that it is in its correct position relative to the cooling water connections, and also that it does not touch the shaft sleeve. The remaining turns of packing can then be inserted, and the gland nipped up until it just enters the stuffing box. No further adjustment must then be attempted until the pump is running as, due to the very high shaft speeds involved, great care must be taken to always allow some leakage through the gland, otherwise overheating will occur with possible damage to packing or shaft sleeve.

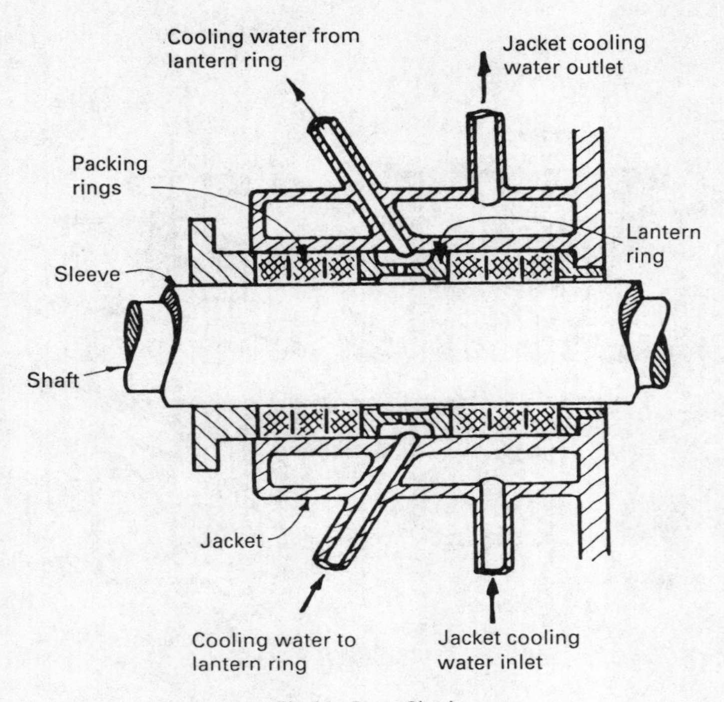

Fig. 34 Pump Gland

With the pump running tighten each gland nut one cant. Allow some minutes to run in and repeat the process as necessary until the leakage has been reduced to the minimum amount required for lubrication and cooling. If the leak off stops, immediately ease off the gland nuts to restore the flow of water.

A by-pass valve is fitted to the pump discharge to prevent overheating by providing some circulation through the pump when it is running against a closed discharge valve, or at low load conditions.

An overspeed trip must be fitted to prevent damage in the event of the turbine overspeeding if the pump loses its suction.

The trip shown in Fig. 35 consists of an eccentric ring mounted at the turbine end of the shaft, and held concentric to it by means of a spring. In the event of overspeed the out of balance centrifugal force set up on the ring causes it to overcome the spring compression, and fly off centre. It then strikes the trigger mechanism so releasing the pawl holding the spring loaded steam valve in the open position. With the pawl released, the spring causes the valve to close so shutting off the steam supply to the turbine and stopping the pump. This mechanism can also be tripped by hand.

To reset, the valve handwheel is turned to the hand closed position so pushing the spring loaded sleeve back into the valve bridge. The pawl can then be reset to hold the sleeve in this position and the handwheel used to open the valve in the normal manner.

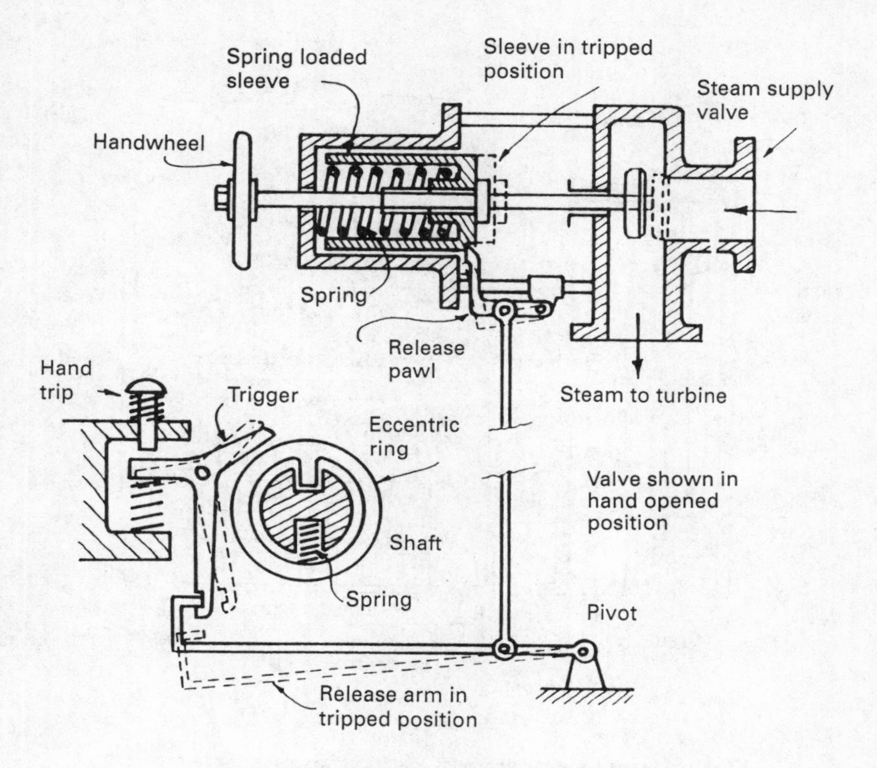

Fig. 35 Overspeed Trip, Spring Operated Valve

A hydraulic governor may be fitted to provide for more economic operation of the pump by controlling the steam supply to suit varying load conditions; see Fig. 42.

TURBO-WATER LUBRICATED PUMP

This consists of a two stage, hydraulically balanced centrifugal pump driven by a Curtis type turbine. The rotating elements are mounted on a common shaft which is supported by two water lubricated bearings, the whole being totally enclosed in a single casing, thus eliminating the need for steam and water glands.

The general arrangement of the pump is shown in Fig. 36. The turbine operates at about 7500 rpm with steam pressures of up to 7000 kN/m^2. Attachment of the turbine wheel to the shaft is by means of a Hirth coupling. This only requires a single bolt, so doing away with the need for a large hub; it also gives good allowance for expansion, while at the same time ensuring correct alignment and helping to reduce the amount of heat conducted along the shaft. This latter effect is due to the hollowed out portion in way of the coupling, which forms a heat barrier between the turbine wheel and the bearings.

A relief valve fitted to the turbine exhaust belt is set to lift at about

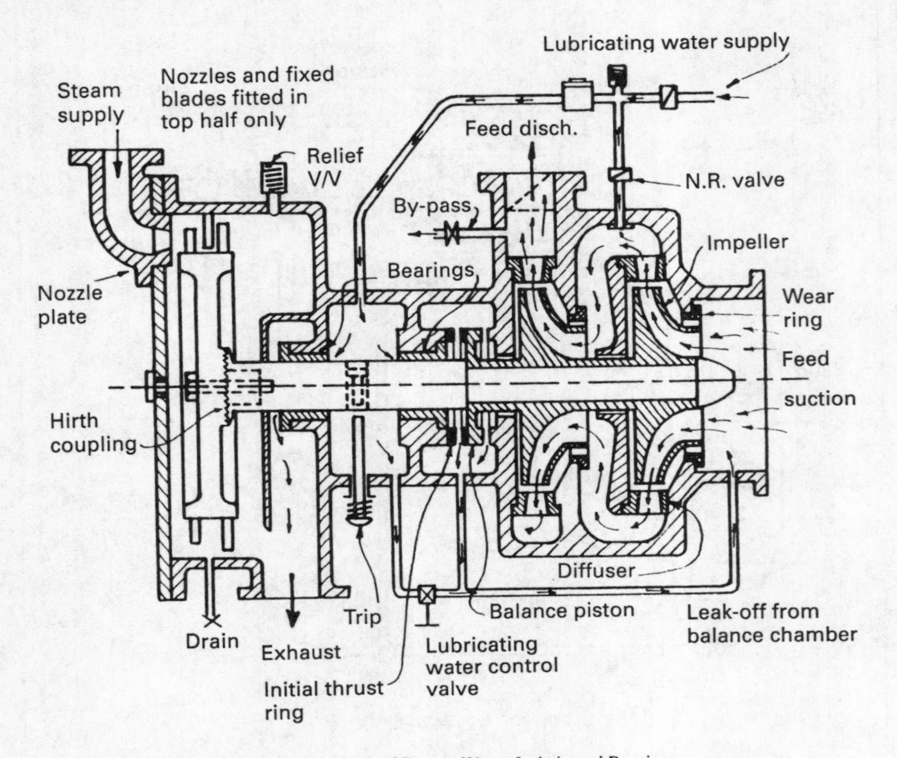

Fig. 36 Turbo-Feed Pump, Water Lubricated Bearings

150 kN/m^2 above the normal exhaust pressure of 300 kN/m^2.

Out of balance forces set up by disturbances to the hydraulic balance arrangements when starting or stopping, are absorbed by an internal initial thrust ring. This is faced with a special material, called Ferrobestos, to reduce wear.

The water lubricated bearings consist of a steel backing onto which is sintered a layer consisting of porous bronze impregnated with lead powder and polytetrafluoroethylene, the latter usually being referred to by the initials PTFE. This plastic forms a thin layer about 0·025 mm thick on the bearing surface. In service some of this is transferred to the shaft so giving a very low coefficient of friction.

Lubricating water is supplied to the bearings from either the extraction pump, or the first stage discharge. In this latter case provision must be made for an alternative supply when starting or stopping the feed pump. The lubricating water having passed through the bearings, returns either to the pump suction, or into the turbine exhaust.

The bearings operate at about 115°C with a lubricating water supply at a pressure of 550 kN/m^2 and temperature of 70°C at full load condition. Bearing clearances when new are 0·15 mm, being renewed at about 0·25 mm. Under no account must this clearance be allowed to exceed 0·3 mm. In practice a useful guide to bearing wear can be obtained by examining the bearing surface with a magnifying glass. If the loaded area shows less than 25% of PTFE, the bearing should be renewed.

As the bearing clearance increases, the water control valve must be adjusted to maintain the correct pressure in the hydraulic balance chamber.

It is essential that the lubricating water is supplied to the bearings before starting up the pump and left on for some time after the pump has been stopped, in order to cool the bearings and prevent them from being overheated by residual heat from the turbine wheel.

An overspeed trip is fitted which operates at 15% overspeed, shutting off the steam supply to the turbine. The arrangement of this is shown in Fig. 37.

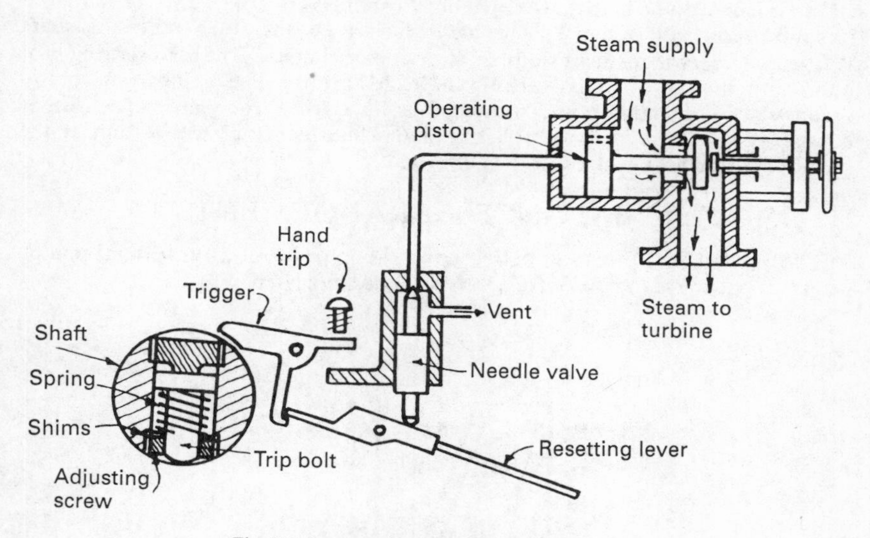

Fig. 37 Overspeed Trip, Steam Operated Valve

The trip consists of an eccentrically loaded bolt mounted in a hole through the shaft and held concentric to it by a spring. In the event of overspeed, the out of balance centrifugal force set up overcomes the compression of the spring and the bolt flies out and strikes the trigger mechanism. This in turn causes the needle valve to open so releasing steam from one side of the steam stop valve actuating piston, closing the valve and shutting off the steam supply to the turbine.

Unlike the previous overspeed trip considered in Fig. 35 this arrangement can be adjusted by inserting shims between the adjusting plug and the end of the spring as shown in Fig. 37. Reducing the total shim thickness reduces the tripping speed, each shim of 0·15 mm thickness altering the trip speed by about 100 rpm.

A by-pass valve is fitted to prevent overheating by enabling some circulation through the pump to be provided when it is running against a closed discharge valve, or at low load conditions.

Some form of governor will be fitted to regulate the steam supply to suit the load condition and so provide more economic operation. This governor may be of the type shown in Fig. 42 but in many cases a more sensitive differential pressure type is fitted.

When the pump is required to act as a standby unit, an automatic cut-in device may be fitted to ensure continuity of supply. In this case a spring loaded non-return valve must be fitted to the pump discharge to ensure that when acting as standby unit with suction and discharge valves open, water is not pumped back from the running pump. This type of valve may also be fitted to make sure the pump quickly builds up pressure upon starting to rapidly establish its hydraulic balance.

The casing is made of cast steel, with a nozzle plate of creep resistant steel. The stainless iron turbine blades are fitted onto a stainless steel wheel which in turn is mounted at one end of a nickel chrome steel shaft. At the other end of this shaft is fitted an impeller made of stainless steel or monel metal. The diffuser ring is of aluminium bronze, and the wear rings of leaded bronze. The oil lubricated pump is constructed of similar materials except for its bearings and glands. The former consist of white metal on a steel backing, while the latter use carbon at the turbine end and soft packing at the pump end.

HYDRAULIC BALANCE ARRANGEMENTS

Due to the unequal pressures acting upon the impeller of a centrifugal pump, powerful out of balance axial forces are set up as indicated in Fig. 38.

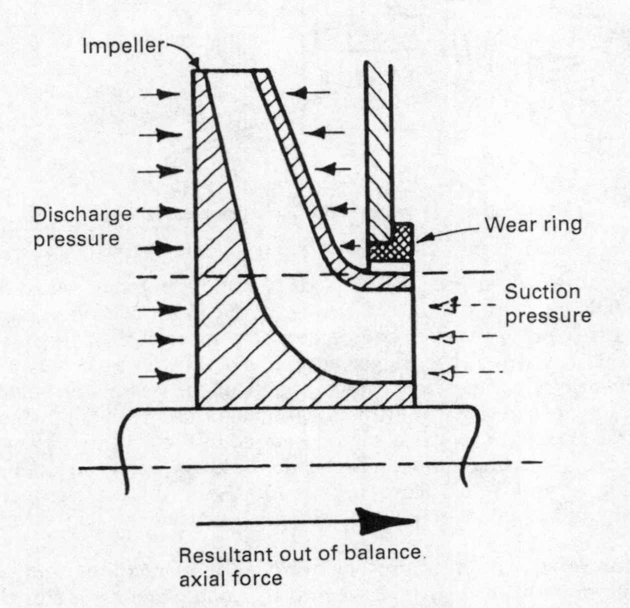

Fig. 38 Hydraulic Forces Acting on Centrifugal Impeller

To reduce the magnitude of the resultant of these forces it is common practice to hydraulically balance centrifugal pumps, so reducing the requirements of any thrust blocks or rings which may be fitted. In turbine driven pumps the axial thrust on the turbine blades can also be used to offset some of this force.

BALANCING FOR SINGLE STAGE

As shown in Fig. 39 a controlled leakage of water is allowed to take place from the pump discharge into the balance chamber. A control opening then permits some of this water to pass through to the leak off chamber, from whence it is usually returned to the pump suction via a balance connection.

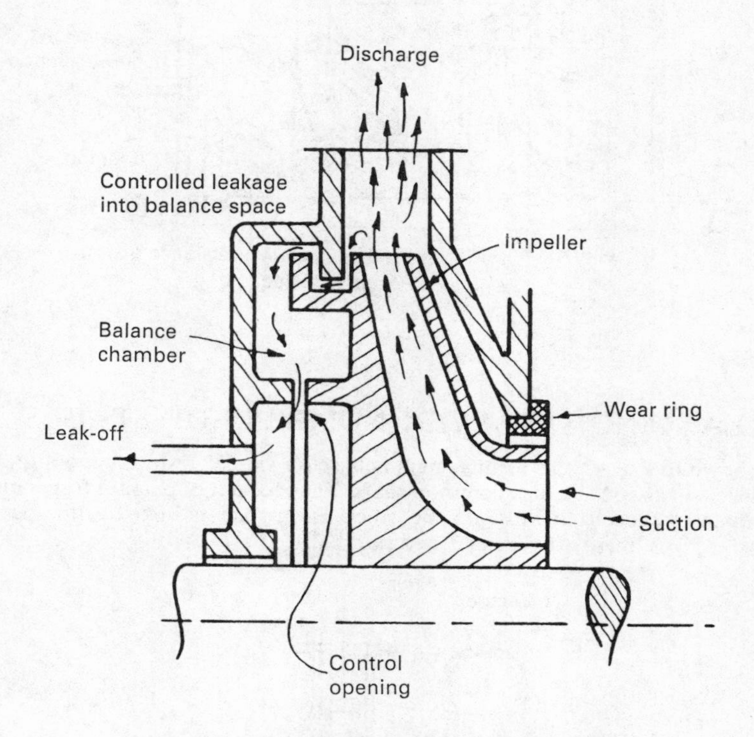

Fig. 39 Hydraulic Balance for Single Stage Pump

An increase of pressure in the balance chamber tends to move the impeller towards the suction side so increasing the control opening which, by allowing more water to leak off, reduces the pressure in the balance chamber. This allows the impeller to move back towards its original position so closing in the control opening and causing the pressure to again build up. This course of events maintains the axial location of the rotating elements of the pump within fine limits, even with varying load conditions.

BALANCING FOR MULTIPLE STAGES

The arrangement shown in Fig. 40, while operating basically in the manner as that previously described, now makes use of a balance piston. The variable control opening again provides the hydraulic balance to maintain the axial position of the rotating elements.

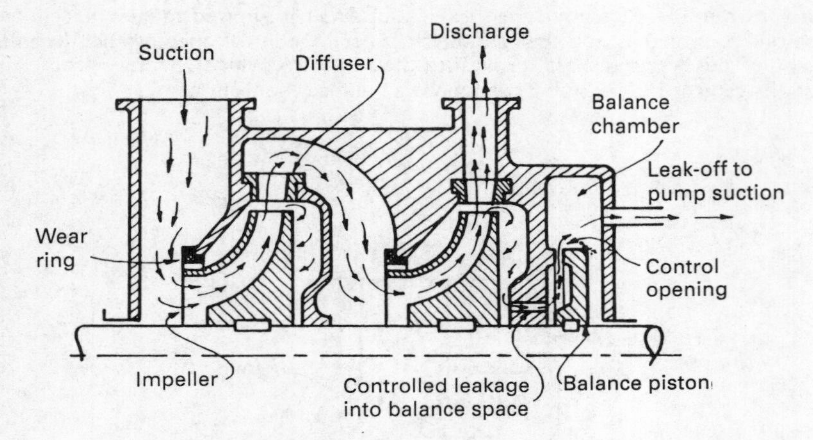

Fig. 40 Hydraulic Balance for Multi-Stage Pump

BALANCING FOR EVEN NUMBERS OF STAGES

As shown in Fig. 41 this arrangement achieves a similar result by mounting the impellers back to back on a common shaft. Thus the out of balance forces on the one help to cancel out those on the other. Each stage is linked to the next by transfer pipes, these often being fitted externally.

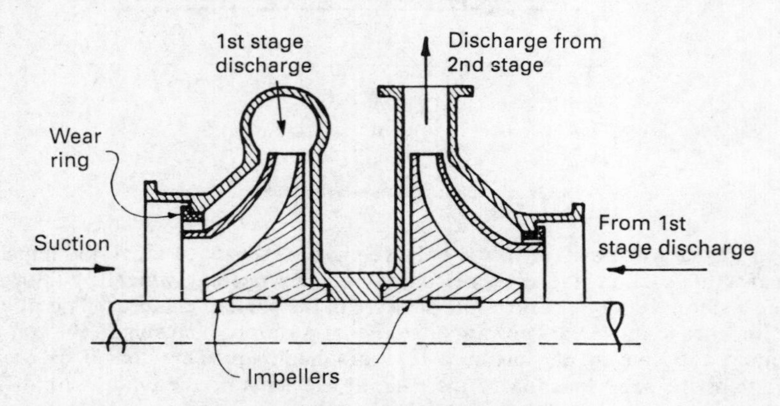

Fig. 41 Hydraulic Balance by Opposed Impellers

HYDRAULIC GOVERNOR

These are fitted to reduce the steam consumption of a turbo-feed pump by regulating the supply of steam to the turbine in response to changes in load and, by providing a constant supply pressure to the feed regulator, ensure a more stable control over the water level.

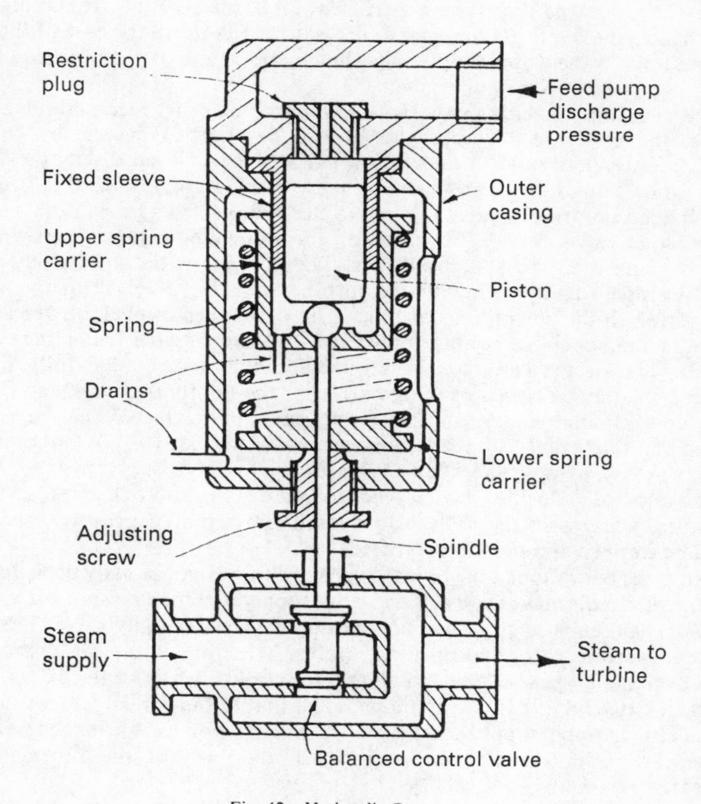

Fig. 42 Hydraulic Governor

In the governor shown in Fig. 42 the feed discharge pressure acts upon the top of the piston pushing it down against the spring. Thus when the boiler feed regulator closes in reducing feed demand, it causes the feed pressure to increase and the piston moves down, acting on the top of the valve spindle to close in the balanced throttle valve. This reduces the steam supply to the turbine which in turn lowers the feed pressure. When this happens the spring causes the piston to move upwards again, opening the throttle so admitting more steam to the turbine and increasing the feed discharge pressure. In this way the hydraulic governor attempts to maintain a constant feed pressure in spite of load variations.

The normal working feed pressure can be altered by means of the adjusting nut fitted to the governor; when this is screwed upwards it increases the spring compression so raising the working feed pressure.

TURBO-FEED PUMP OPERATION

The general starting procedure is, first to check that the pump is in good order and free to rotate and that there is sufficient oil in the bearings. If cooling water is supplied to the pump glands, make sure it is turned on. In the case of the water lubricated pump the water supply must be on and at the correct pressure before attempting to start up the pump.

Make sure pump isolating valve is closed, and if no other feed pump is already on line, the boiler feed checks should also be shut. Hydraulically balanced centrifugal pumps should always be started against a closed discharge valve to ensure a rapid build up of pressure. In many cases, to make sure this is done, a spring loaded non-return valve is fitted to the pump discharge. This also enables the discharge valve to be left open ready for a quick emergency start when the pump is being used for standby duties. The pump suction and by-pass valves should be opened together with the turbine exhaust valve. Then the various drains fitted to the steam line, turbine casing and carbon packing are opened, before cracking open the steam supply valve and allowing the steam line to warm through. The drains can then be closed and the steam valve fully opened. Provided the pump is then running correctly, the pump discharge valve can be slowly opened if necessary, and the pump put into service. When the pump is running at a fairly constant load above a value of at least 10% of full output, or at full away, the by-pass valve can be closed.

If at any time during this procedure undue vibration is experienced, or lubricating water pressure falls below minimum required pressure, the pump should be stopped and the fault rectified.

To stop the pump, open the by-pass valve if this has not already been done and close the pump discharge valve. The steam supply can then be shut off either by means of the manual trip, or by hand closing the steam supply valve. Then close the exhaust steam valve and open the drains. When the pump has come to rest the suction and by-pass can be closed and the cooling water to the gland cooling jackets, if fitted, shut off. It should be noted that in the case of water lubricated bearings, their water supply should be left on for some time after the pump has stopped, in order to cool the unit down and prevent the bearings being overheated.

Various routine test procedures are necessary; two of the most important of these, which should be carried out at least once every six months, are given here. The first is that for the overspeed trip. Make sure the manual trip device is operating corectly and that the turbine speed can be measured accurately; this is usually done by fitting a tachometer to the end of the turbine shaft. Then with the pump running against a closed discharge valve but with the by-pass valve open, close in the suction valve until either the overspeed trip operates or the tachometer indicates that the pump has reached its maximum allowable speed. In this latter case immediately use the manual trip to stop the pump. As soon as the trip has operated, quickly restore the pump suction by opening the suction valve. The tripping speed is usually typed on one of the cover plates fitted to the casing.

The second test is for the pressure relief valve on the turbine casing. In this case the pump is stopped with all valves, except for the casing drain, closed. The steam supply valve is then cracked open until some steam issues from the open drain, which is then shut in, causing the steam pressure in the casing to rise. If the relief valve does not open as required, reopen the drain to release the pressure, then reset the relief valve and retest. When the valve lifts at the correct pressure, replace and repin its protective dome.

In addition to these tests the axial clearances should be checked at regular and frequent intervals, and adjustment carried out as necessary.

ELECTRO-FEEDERS

These consist of multi-stage centrifugal pumps, hydraulically balanced and driven by a constant speed electric motor. Their normal running speed will be in the order of 3000 rpm and as this is much lower than that of turbine driven pumps for similar discharge pressures, electro-feeders require a greater number of stages. Depending upon the quantity and discharge pressure needed the number of stages varies between two and ten. The intermediate stages consist of identical sections, located by spigots, the joints sealed by 'O' rings and then clamped together between the inlet and discharge sections by a number of external tie bolts. Each stage consists of a fixed ring, containing a vane diffuser and incorporating a diaphragm to separate the individual pump stages. Special sealing arrangements are provided in way of the shaft to reduce interstage leakage.

As shown in Fig. 43 the impellers are keyed onto a common shaft which is supported at each end. The impeller hubs are extended to abut each other and a locking strip is then used to firmly clamp these impellers together with the hydraulic balance piston, between two split rings which fit into circumferential grooves in the shaft.

In the version shown, a water lubricated PTFE bearing is used to support the shaft at the inlet end, similar to those of the TWL pump, except that no separate lubricating water supply is required, the water passing through the electro-feeder itself serving this purpose. At the other end of the shaft is fitted a thrust ball race which not only provides support but also locates the rotating elements and deals with any small residual thrust left by the hydraulic balance arrangements.

No gland need be fitted at the inlet end of the pump but at the discharge end where the shaft passes through the pump casing a mechanical seal is fitted.

In other versions of these electrically driven pumps the shaft is supported by two external oil lubricated bearings, while the glands fitted at each end of the pump casing use soft packing.

A balance connection is provided to return the leak off water from the hydraulic balance chamber to the pump suction. An air vent is provided on the first stage section for use when putting the pump into service. A spring loaded non-return valve is fitted at the discharge to ensure that the pump starts under light load condition without overloading the motor, and then to ensure a rapid build up of pressure so that the hydraulic balance is quickly established. Finally a by-pass valve is fitted to prevent overheating, by giving some circulation through the pump at low load conditions, or when running against a closed discharge valve.

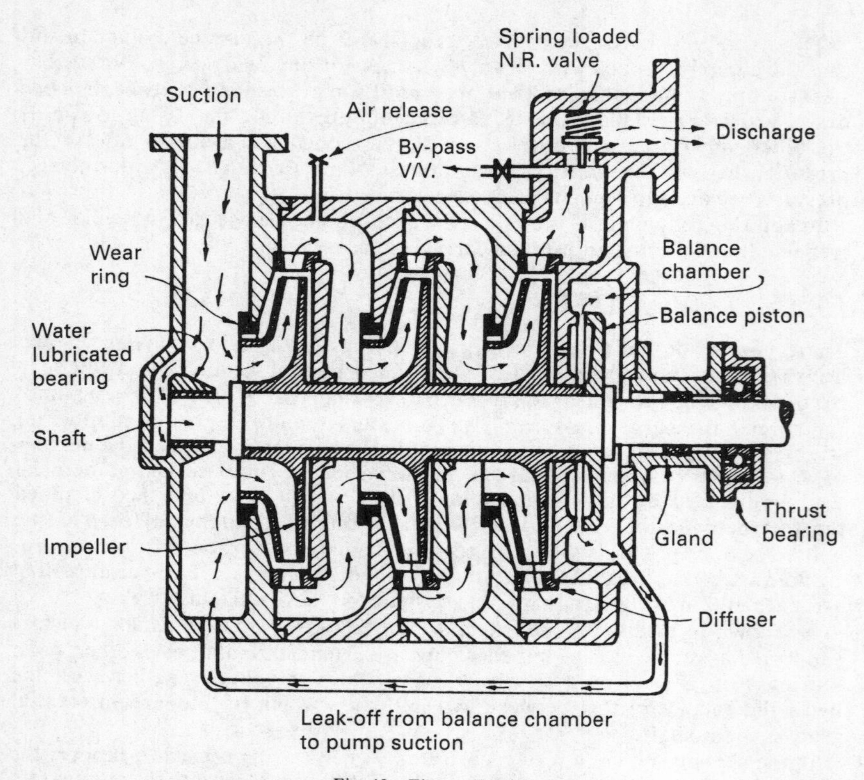

Fig. 43 Electro-Feeder

STARTING PROCEDURE

Check the pump is in good condition and free to turn by hand, and that the motor is wired to give correct direction of rotation. Next open the inlet valve to prime the pump and use the vent on the first stage to release air from the pump, making sure the pump is full of water and the by-pass valve open. Check the discharge valve is closed and, if necessary, the boiler feed check. The motor can now be started and the discharge valve slowly opened. At full away the by-pass valve can be closed.

CHAPTER 8

Feed Water Control Systems

The water level in a boiler is critical. If it is too low damage may result from overheating; if too high priming can occur. It is therefore desirable to maintain the level in the boiler within the limits of the water level indicator.

In the case of water tube boilers with their high evaporation rates and small water reserves, the classification societies demand that some form of automatic feed water control be fitted, to maintain the water level in the boiler drum within the desired limits. These control systems may be grouped into the following types.

SINGLE ELEMENT

These function in response to changes of the water level in the boiler. They use either a float or thermal means to sense the water level and may be self-actuated or use an external power source.

TWO ELEMENT

In addition to the water level, this type of regulator also senses changes in steam demand, and by so doing attempts to reduce the wide fluctuations of water level during manoeuvring which some boilers are prone to. The reason for this is that all water tube boilers are subject to the phenomenon of shrink and swell, resulting from changes in steam demand causing variations in the rate at which steam bubbles form below the water level. A sudden increase in steam demand leads to a slight decrease in the drum steam pressure which, coupled with a higher rate of firing, results in a rapid increase in the rate of bubble formation. This causes the water to swell up into the steam drum. The term shrinkage refers to the opposite effect, where a reduction of bubble formation follows a sudden decrease in steam demand. Thus the shrink and swell characteristics of the boiler, which are largely dependant upon the size of the steam drum relative to the evaporation rate of the boiler, may have a marked effect upon the water level, leading in some cases to considerable fluctuations when manoeuvring.

This in turn can cause the following problem when a single element type feed regulator is fitted. As the water swells up into the drum, the regulator sensing the rise of water level shuts in, thus reducing the flow of feed water into the boiler. When the level falls as steam is consumed, the regulator responds by admitting more feed. This relatively cool water entering the boiler absorbs more heat so reducing the rate of steam bubble formation. The resultant shrinkage lowers the water level causing the regulator to admit even more feed, thus aggravating the effect. This condition, where the level continues to fall in spite of the regulator position, is referred to as hunting of the water level. In many boilers it is slight and no special measures need be taken, but if excessive, one method of reducing it is to fit a two element feed regulator. This, upon sensing an increased steam demand, will cause more feed to be admitted even though the boiler water level is already rising. The two effects thus tend to cancel each other out. Although this type can be self-actuated, most modern versions use an external power source.

THREE ELEMENT

These in addition to the water level also measure the steam and feed flows. The basic control action now functions from the relation between the steam and feed flows. In a closed feed system these will be equal under steady load conditions, any change from this equilibrium condition resulting in a control signal being applied to the feed water control valve, so adjusting the supply of feed to suit the new load condition. The water level then supplies a signal which is used to trim this level back to the desired value when stability is again established.

Regulators of this type invariably use an external power source.

WEIR'S ROBOT FEED REGULATOR

This is a self-actuated, float operated single element type giving proportional control. The basic layout is shown in diagrammatic form in Fig. 44.

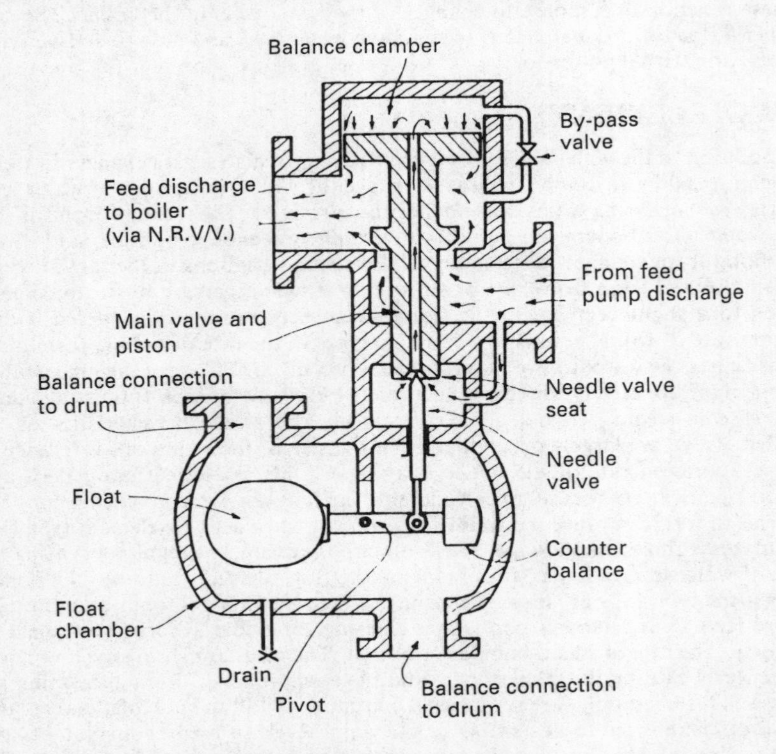

Fig. 44 Float Operated Feed Regulator, Weir's Robot Type

Boiler pressure is acting upon the underside of the actuating piston and on the top of the valve lid, while the feed discharge pressure acts on the underside of the valve lid. As the piston area is twice that of the valve lid, and the feed discharge pressure is about 1000 kN/m^2 in excess of the boiler pressure, the net result is an upward force tending to open the valve. This is then balanced by a downward force exerted on the top of the actuating piston due to the pressure set up in the balance chamber by water from the feed discharge passing to it via the needle valve. This valve which is attached to the float arm, controls the pressure by varying the supply of water to the balance chamber in response to changes in the float position. When the pressure in the balance chamber is approximately midway between the feed and boiler pressures, the valve takes up an equilibrium position and so provides a steady flow of feed to the boiler.

When the boiler water level rises it lifts the stainless steel float causing the monel metal needle valve to be lowered, which allows more water to flow to the balance chamber increasing the downward force on the actuating piston. This tends to close in the control valve and reduce the feed flow. When the water level drops it lowers the float, so raising the needle valve and thus reducing the flow of water to the balance chamber. As water is continually leaking past the piston the pressure in the balance chamber now rapidly falls, which allows the valve to open up so increasing the feed flow.

If the steaming rate of the boiler is changed for an extended period, it should be noted that as this type of regulator only gives proportional control, the resulting slight variations of pressure across the valve cause it to take up a new equilibrium position. This leads to a slight change of water level in the boiler for the new load condition. This level can be adjusted by altering the thickness of washers under the needle valve seat. Increasing the washer thickness will increase the water level by the ratio of 1 to 7.

In case of emergency the small by-pass valve can be opened to release the pressure in the balance chamber and allow the valve to come full open. When in service the float chamber should be blown through occasionally to prevent a build up of sediment.

Excessive clearance in way of the actuating piston causes a high water level, while wear in way of the needle valve seat causes a low water level.

MOBRAY FEED REGULATOR

This is a float operated magnetic type, single element feed regulator, with an external electric power source. A diagrammatic layout is shown in Fig. 45.

The regulator shown consists of a cast steel float chamber, suitable for pressures up to 1750 kn/m^2, containing a stainless steel float to which a permanent magnet is attached, An aluminium bronze switch housing mounted on this float chamber contains an air break electric switch with silver contacts, which is operated magnetically through the wall of the switch body. This is effected by means of the permanent magnet carried by the float assembly which is opposed by the similar magnet in the switch housing. The adjacent poles are arranged to repel each other so operating the switch with a snap action. Thus no moving parts, other than the float, are in contact with the water and therefore no glands are required.

This type of regulator provides an on and off control action, suitable for use with an electrically driven feed pump. The switch contacts being wired in con-

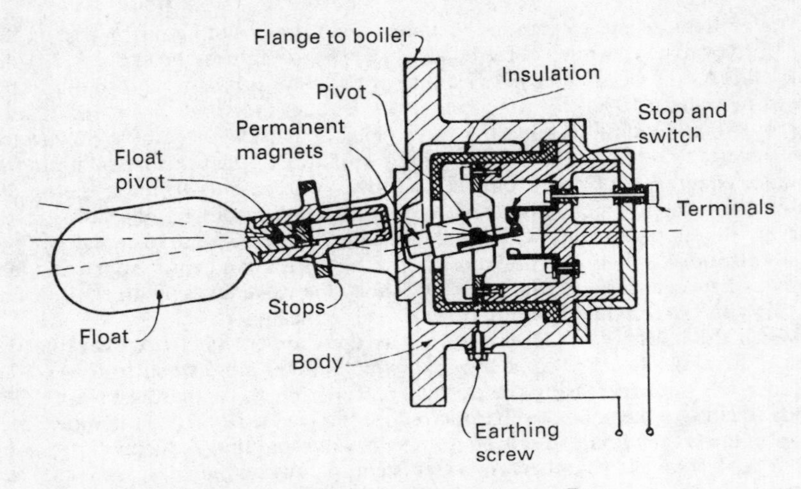

Fig. 45 Magnetic Feed Regulator, Mowbray Type

junction with two make-before-break contacts, and connected in series with the operating coil of the feed pump motor starter. Thus for small auxiliary tank type boilers, where some considerable change of water level can be tolerated, the regulator forms a simple reliable unit.

When the water level falls, the magnet attached to the float moves opposite the switch magnet, which is immediately repelled as the like poles line up, snapping across to close the switch contacts and starting up the feed pump. Water then enters the boiler raising the level until again the float magnet comes opposite the switch magnet, which is again repelled, snapping back to its original position so opening the switch contacts and stopping the pump.

FISHER LEVELTROL

This is a float operated single element regulator, using an external pneumatic power source. The basic layout is shown in diagrammatic form in Fig. 46.

Movement of the float causes the torque arm to move the flapper. Changes of the water level thus cause the gap between the nozzle and the flapper to alter, so varying the air pressure in chamber A. This pressure acts upon the diaphragm which in turn moves exhaust valve C and inlet valve B. This controls the air pressure acting upon the actuating diaphragm for the feed control valve.

As the water level rises, the gap between the nozzle and the flapper decreases so allowing less air to escape and causing a build up of pressure in chamber A. This, acting upon the diaphragm, first closes the exhaust opening, and then lifts valve B off its inlet seat increasing the output pressure to the actuating diaphragm. The feed control valve then closes in reducing the feed flow to the boiler.

When the water level falls the gap is increased allowing more air to escape and reducing the pressure in chamber A. This allows the spring to close valve B and then, as pressure in chamber A continues to fall, movement of the diaphragms allows exhaust valve C to open to the atmosphere. The output pressure on the

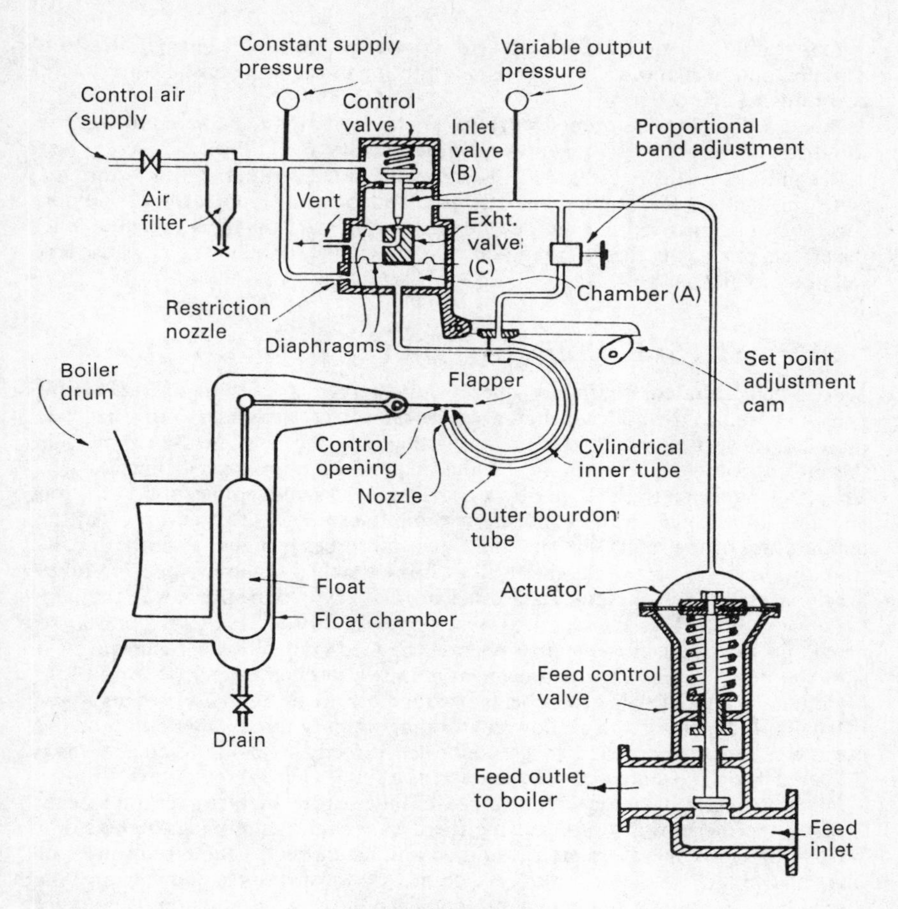

Fig. 46 Fisher Leveltrol

actuating diaphragm is thus reduced, allowing the feed control valve to open under the action of its spring and increase the feed flow into the boiler.

This regulator, although more complicated than the preceding types considered, gives a more stable control action due to its much wider proportional band, which enables the water level to be maintained close to the desired value over a wide load range. Under steady load conditions the feed control valve takes up an equilibrium position so that feed flow equals steam flow. This is achieved by the negative feed back device consisting of the outer bourdon tube which, as the ouput pressure varies, changes its radius so moving the nozzle away from the flapper. This stops the pressure change taking place in chamber A and allows the valve to take up a new equilibrium position to suit the new steady load condition. There will however be a slight change in the position of the new water level.

This regulator being of single element type cannot anticipate changes in steam demand, and so cannot effectively reduce hunting of the water level if this should occur during manoeuvring.

It is essential for the proper working of the regulator that the control air is supplied at constant pressure, and kept clean and dry at all times.

In some cases a displacer is used instead of a float. These sink partway into the water instead of floating on the surface, and are thus subjected to varying degrees of upthrust as the water level changes. They are thus less susceptable to the effects of surging than a float, and also, by giving a smaller travel, cause less wear on moving parts.

COPES SINGLE ELEMENT FEED REGULATOR

This is self-actuated, thermally operated type regulator, with its thermostat fitted external to the boiler, being mounted on a fixed triangular frame which is placed in a vertical plane and in line with the water level in the drum. The thermostat consists of two similar inclined expansion tubes, each inclined at 45°. They are connected to the boiler by a heavily lagged steam connection at the top end, and an unlagged water connection at the bottom end. This ensures that the temperature of the steam entering the expansion tubes is higher than that of the water in the lower part of the thermostat. Lugs attached to these expansion tubes are connected to a bell crank via a series of links pivoted in such a way that any movement of the lugs is magnified by a factor of 70 to 1 at the bell crank, so providing sufficient movement to operate the feed control valve by means of a suitable rod or chain. This arrangement is shown in diagrammatic form in Fig. 47, the same type of thermostat being used in both single and two element types. If the diaphragm attachment shown in the diagram is ignored, and point B on the operating rod is considered to be connected directly to point C on the valve operating lever, the unit will then function as a single element regulator.

When the boiler water level falls it causes more of the high temperature steam to enter the thermostat tubes causing them to expand, so moving the lugs outwards. This movement is then transmitted and magnified by the arrangement of links and pivots, to the bell crank which moves downwards as indicated by the dotted lines in Fig. 47, so opening up the feed control valve and admitting more water to the boiler.

When the water level rises, more relatively low temperature water entering the thermostat tubes causes them to contract, thus moving the links and rods in the opposite direction so closing in the feed control valve and reducing the flow of water to the boiler.

Adjustment of the water level can be made by means of a turnbuckle on the valve operating rod. The thermostat is comprised of two tubes so that when the ship heels, water rises in one leg and falls in the other, these effects cancelling each other out at the bell crank.

The feed control valve is balanced, opening or closing ports in a cage. A spring is fitted to this valve which tends to return it to the full open position.

TWO ELEMENT REGULATORS

These are usually fitted in order to reduce hunting of the water level.

COPES TWO ELEMENT FEED REGULATOR

This is a self-actuated type, having a similar thermostat arrangement to measure the water level as the single element type previously described, but also includes a spring loaded diaphragm in the system which enables changes in steam demand to be measured. The basic layout is shown in Fig. 47.

The upper side of the diaphragm is connected to the superheater outlet, while the lower side is connected to the superheater inlet. In both cases the connections are made by unlagged pipes; this is so that steam condensing in these pipes provides a water shield protecting the diaphragm from the high temperature steam.

An increase in steam demand reduces the pressure at the superheater outlet so causing an upward movement of the diaphragm. This movement is then conveyed through a series of links to the valve operating lever, opening the feed control valve and increasing the flow of water into the boiler. A reduction in steam demand will cause the superheater outlet pressure to increase, thus resulting in a downward movement of the diaphragm. This causes the links to move in the opposite direction so closing in the feed control valve and reducing the flow of water into the boiler.

Thus the feed control valve is operated by either a change in the water level effecting the thermostat or by a change in steam load effecting the diaphragm.

Adjustment to the amount of control exerted by the diaphragm movement can be made by moving the point of attachment of the diaphragm linkage onto the valve operating lever.

The diaphragm is not designed to withstand full boiler pressure, thus a cross connection is fitted which balances the pressures across the diaphragm when the connections to the superheater are being opened or closed. The cross connection only being closed when the valves in both the superheat connections are full open.

After renewing the diaphragm it is advisable to fill the connections to the superheater with distilled water, so making sure the high temperature superheated steam is kept away from it. Condensing steam keeps these unlagged connections filled with water when in service.

COPES PNEUMATIC TWO ELEMENT REGULATOR

This is a two element type using an external power source. It has two control elements, these consisting of a thermostat responsive to changes in the water level, and a steam flow element responsive to changes in steam demand. The basic layout is shown in diagrammatic form in Fig. 48.

The thermostat functions on the same basic principle as that of the self-actuated two element type previously described, but differs in detail as only a single inclined stainless steel tube is used. The upper end of this tube is attached directly to a rigid frame which is mounted external to the boiler and in line with the drum water level. The lower end of the tube is free to move. The thermostat lever and a bell crank are then used to transmit and magnify any movement of the expansion tube, so as to operate a pilot valve which controls the air pressure supplied to diaphragm A.

The steam flow element consists of a spring loaded reinforced diaphragm B, with connections to the superheater inlet and outlet. These connections are

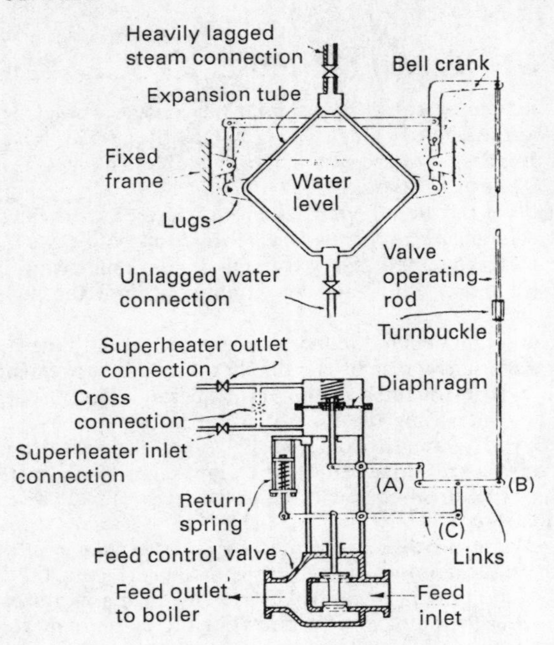

Fig. 47 Cope's Two Element Feed Regulator

unlagged so that condensing steam provides a water shield to protect the diaphragm from the high temperature steam. A cross connection is fitted to prevent full boiler pressure being applied to only one side of the diaphragm when opening or closing the isolating valves in the superheat connections. This cross connection is only closed when both these isolating valves are fully opened.

Movement of the two diaphragms A and B is combined and transmitted through a system of links and fulcrums to operate another pilot valve. This then regulates the air supply to the actuator for the feed control valve.

When the steam demand increases, the pressure drop across the superheater also increases, resulting in changes in the pressure acting upon diaphragm B so causing it to move. Levers C and D then transmit this movement to operate the pilot valve regulating the air supply to the actuator of the feed control valve. This opens the valve to admit more water to the boiler. This in turn provides the primary control signal in the system and endeavours to regulate the inflow of feed so that it equals the outflow of steam. Due however to the time taken for the regulator to respond to any changes in flow conditions it is also necessary to measure the feed level and this is done by the thermostat. Thus if the water level rises, more relatively cool feed water enters the thermostat tube causing it to contract. This movement operates the associated pilot valve which regulates the supply of air to diaphragm A causing it to move. Lever D then moves so operating the pilot valve regulating the air supply to the valve actuator, which in this case results in the feed control valve closing in and reducing the flow of water into the boiler.

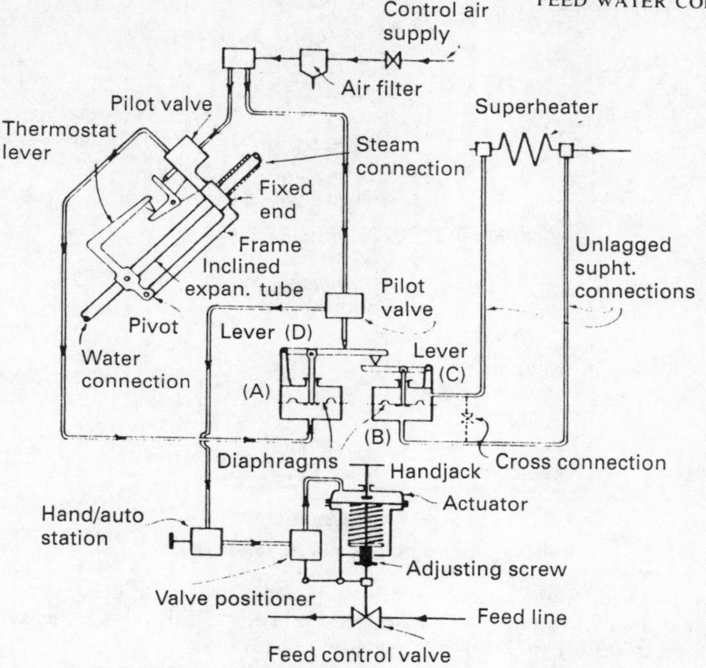

Fig. 48 Cope's Pneumatic Type Feed Regulator

AUTOMATED TWO ELEMENT REGULATOR

This is a two element regulator using an external power source. One of the two conditions measured is the variation of steam flow caused by changes in steam demand. The signal obtained initiates the primary control action for the system, which functions to open or close the feed control valve to suit changing steaming rates. Theoretically this control action could be arranged so that the inflow of feed to the boiler would equal the outflow of steam. In practice however due to various time lags in the system it is also necessary to measure the feed level; changes in this produce a second signal which is then used to modify the primary control signal in order to avoid undue fluctuations in the water level. The basic layout in the form of a block diagram is shown in Fig. 49.

The mass flow of steam is measured by the pressure drop across an orifice in the steam line; this however gives a non-linear relationship and a square root extractor is included in the system so that the output signal to the relay remains proportional over a wider range of steam flow.

The second condition measured is the water level, which is done by using a differential pressure cell to measure the difference between the water level in the drum and that in the constant head device. As this measurement only involves linear terms, no correcting device is necessary for the transmitter to produce an output signal proportional to any changes between these levels. This signal goes to the automatic controller where it is compared with a predetermined set point signal. Any error between the two signals results in this controller producing a

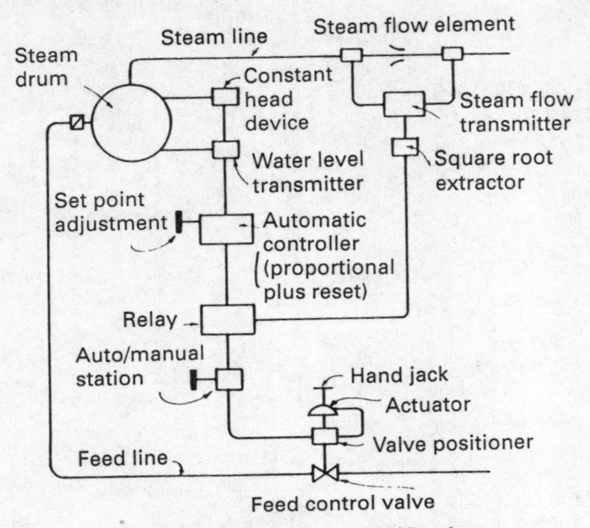

Fig. 49 Automated Two Element Feed Regulator

signal. For sudden or large fluctuations in steam demand a proportional signal only may be inadequate and the addition of an integral, or reset signal will be necessary to restore the level to the set point condition.

The resulting signal from the controller goes to the relay where it adjusts the signal from the steam flow transmitter in order to return the water level to its desired value as given by the set point signal. The signal from the relay then goes to the auto-manual station and thence via the valve positioner to the actuator, which then drives the feed control valve to the position necessary to give the correct flow of water into the boiler.

It is important for stable control that a governor be fitted to the feed pump which enables the feed supply pressure to the control valve to remain approximately constant.

Changes in the operating water level can be made by means of the set point adjustment.

Suitable arrangements must be made for a supply of clean dry control air, at a constant pressure of about 200 kN/m^2.

AUTOMATED THREE ELEMENT REGULATOR

This type uses an external power source, and measures steam and feed flows in addition to the drum water level.

The flow sensors consist of orifice plates or flow nozzles fitted in the pipe lines so that changes in the flow rates cause a variation in the resultant pressure drop across these sensors which can be used to produce a signal at the respective flow transmitters. To keep these signals proportional to a wider range of flow changes, square root extractors are included in the system. The two flow signals produced are then compared to each other in a differential relay, with any difference between them resulting in the production of a primary flow control signal. Theoretically this should provide full control, holding the steam and feed flows equal to each other. However in practice, due to various time lags in the system, and to water losses from the boiler not included in the steam flow such as blow down etc, a water level signal is also necessary to modify this primary flow signal.

The water level is measured, usually by some form of constant head device operating a differential pressure cell, resulting in a signal from the level transmitter. As this measurement only involves linear terms, no square root extractor is needed to keep the signal proportional to the change in water level. The signal goes to the automatic controller where it is compared with a predetermined set point signal. Any error between these two signals results in the automatic controller producing a basic control signal. The controller includes integral control action to cope with changes in steam demand. The resulting signal is then used to modify the primary flow signal in the summating relay.

The final signal from the relay then goes to the auto-manual station and thence via the valve positioner to the actuator which then drives the feed control valve to the position necessary to give the correct flow of feed into the boiler in order to keep the water level within the desired limits. The valve positioner is fitted to ensure that the final valve position corresponds to that demanded by the control signal.

Changes in the operating water level can be made by adjusting the set point signal. In the event of a malfunction of the control system a hand jack is fitted to the feed control valve to provide for direct manual control.

For optimum control action the feed pump discharge pressure should be regulated so as to maintain a constant pressure drop across the feed control valve. One method of doing this is to use the feed control signal to also control a valve regulating the steam supply to a turbo-feed pump so enabling the pump output to keep in step with any changes in feed demand.

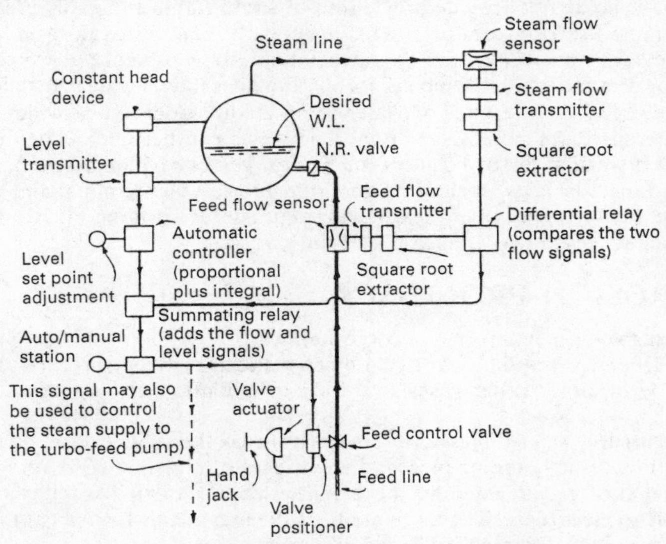

Fig. 50 Automated Three Element Regulator

CHAPTER 9

Feed and Boiler Water Treatment

Corrosion and the formation of scale can lead to serious problems in boilers and feed systems. The corrosive action will waste the metal away thus reducing its overall strength and leading to eventual failure. Scale forming on heat exchange surfaces results in the formation of an insulating layer which reduces the rate of heat transfer. This not only lowers the boiler efficiency, but what is more important also causes the temperature of the metal to rise so reducing its tensile strength. In the case of fired pressure vessels such as boilers, where it is in a highly stressed condition due to steam pressure and thermal effects, this reduction in strength can lead to distortion and again to eventual failure.

Both corrosion and scale are caused by solids and gases dissolved in the boiler water, usually introduced in the feed water supplied.

CORROSION

Metals obtained from the corresponding metallic oxide tend to revert to their original state due to corrosion. Some of these metallic oxides will then form a stable coating which adheres firmly to the surface of the parent metal protecting it from further attack. Other oxides however form coatings which are either porous and so do not provide protection, or are unstable and easily break down. For example a special form of iron oxide known as magnitite can form in boilers. This provides a protective film on the metal but in the presence of oxygen breaks down to form ferric oxide which is porous and does not provide protection.

In the boiler the presence of water in an acidic condition provides the electrolyte required for corrosive action. The water can become acidic due to the presence of various dissolved gases, such as oxygen or carbon dioxide, or various dissolved metallic salts, such as magnesium chloride. Also animal and vegetable fats can break down when heated, the animal fat forming stearic acid and glycerine, the vegetable fat olearic acid and glycerine.

CORROSION PROCESSES

There are two principal forms of corrosion.

Direct Chemical Attack. This can occur when metal at high temperature comes into contact with air or other gases, resulting in oxidation or sulphidation of the metal.

Electro-Chemical Action. This term covers most of the other forms of corrosive attack, those usually taking place in the presence of moisture. One very common form is that of galvanic action, this being set up when two dissimilar metals are placed in an electrolyte. The more noble of the two metals form a cathode to the base metal which, forming the anode, is wasted away.

Other common forms of electro-chemical action occurring in the boiler are hydrogen evolution, causing general wastage of the boiler metal; oxygen absorption, a form of attack which leads to pitting of the metal surface.

IONIZATION

To have a basic understanding of corrosion processes it is necessary to have some knowledge of atomic theory. In very simple terms this can be considered in the following manner.

An atom consists of a number of electrons moving in orbits around a nucleus formed mainly of protons and neutrons. The protons have a positive electrical charge while the electrons have a negative charge. The neutrons, which consist of protons and electrons so close together that their opposite charges cancel, are thus neutral and here may now be neglected.

An atom of any particular substance contains definite but equal numbers of protons and electrons, thus their electrical charges balance and the atom is electrically neutral. It can however then lose or gain electrons, and so obtain a positive or a negative electrical charge. When in this state it is referred to as an ion.

According to the Bohr theory the electron orbits fall into groups, each group or shell having a maximum number of electrons it can hold, two electrons in the first shell, eight electrons in the second, eighteen in the third, and so on. The term valency is used here to indicate the number of electrons required to bring the shells to their maximum number of electrons. Consider the atoms of hydrogen and oxygen as shown in Fig. 51.

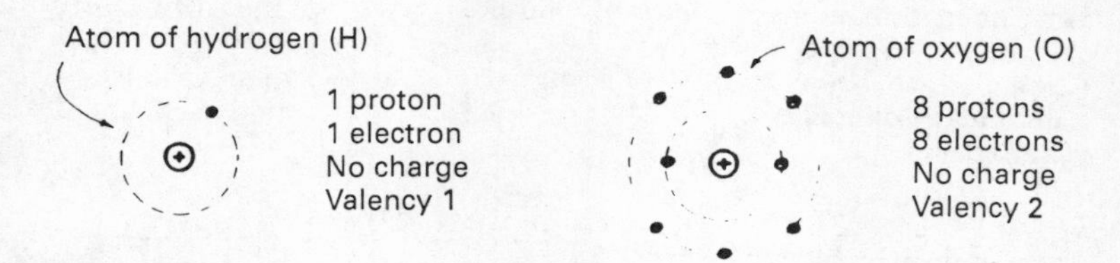

Fig. 51 Conventional Representation of Atoms

The hydrogen atom can contain another electron, the oxygen two more. Thus both atoms have a tendency to acquire more electrons to bring their valency values to zero. A common way of doing this is for hydrogen and oxygen atoms to combine to form a molecule of water. As shown in Fig. 52 the oxygen atom has taken two electrons into its outer ring, which now has its maximum number of eight. The hydrogen shares these, so obtaining its maximum value of two. The water molecule is thus stable, having no electrical charge and a zero valency. But some of these water molecules can become ionized by losing a hydrogen nucleus as shown in Fig. 52.

The hydrogen ions with a positive electrical charge will attempt to stabilize by acquiring electrons from a cathodic surface, for example the boiler metal if the water is allowed to become acidic.

Pure distilled water contains equal numbers of hydrogen ions, and hydroxyl ions, and is thus electrically neutral. However impurities in the water change this rate of ionization, leading to an excess of either the hydrogen or hydroxyl ion

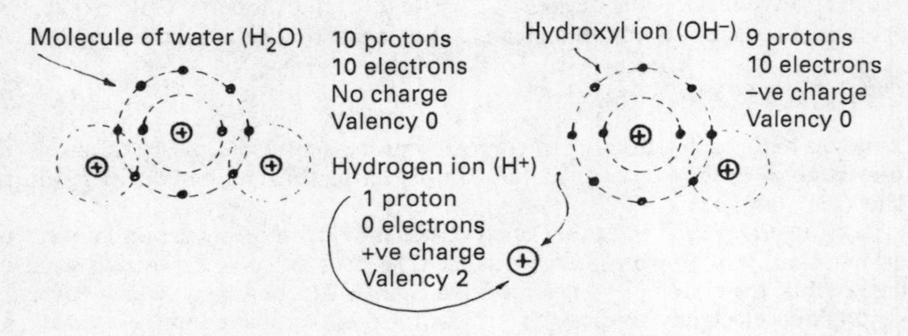

Fig. 52 Conventional Representation of Ions

concentrations. If an excess of hydrogen ions occurs the water becomes acidic, and attack on the metal surface in contact with the water can take place. A typical form of attack known as hydrogen evolution which causes general wastage of the metal surface is shown in Fig. 53. Even in the absence of dissolved oxygen this form of corrosion can take place when the water has pH values below 6·5.

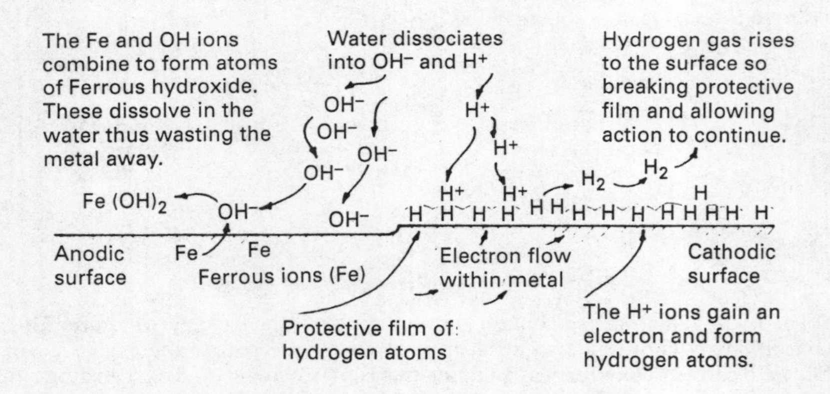

Fig. 53 Hydrogen Evolution Type Corrosion

The anodic surface shown in the diagram is constantly changing position, hence attack occurs over a wide area leading to general wastage of the metal. If dissolved oxygen is present in the water it combines with the hydrogen atoms forming at the metal surface to form water, thus speeding up the process by removing the protective hydrogen layer from the metal more rapidly.

Another form of corrosion known as oxygen absorption is shown in Fig. 54. This can take place when the metal surface is in contact with water having a pH value between 6 and 10, and containing dissolved oxygen. It can cause serious pitting of the metal surface.

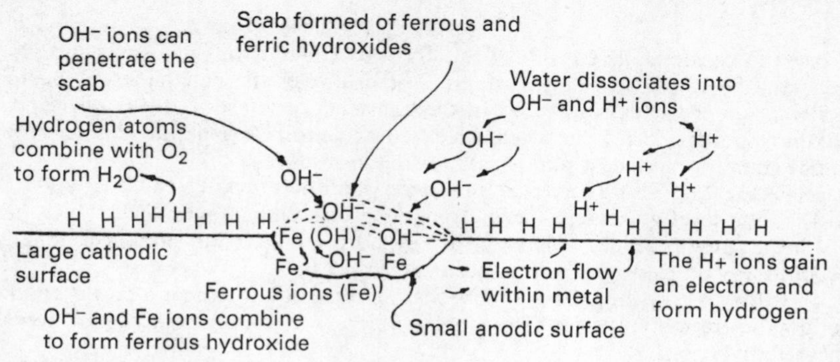

Fig. 54 Oxygen Absorption Type Corrosion

This is a troublesome form of corrosion that can occur if incorrect feed treatment is being used which, while preventing general corrosion, is not adequate to give complete protection. It often takes place in idle boilers. Once started this form of attack cannot be stopped until the rust cap over the pit has been moved, either mechanically by wire brushing, or by acid cleaning. The presence of chorides although a neutral salt often increases this form of attack.

One special form of oxygen absorption corrosion is referred to as deposit attack, and results in local pitting taking place beneath various deposits. It is due to differential aeration, the metal under the deposit becoming anodic in respect to exposed metal which, receiving a more plentiful oxygen supply, becomes cathodic. This form of attack is more common in horizontal than vertical tubes, often being associated with condenser tubes.

FORMS OF CORROSIVE ATTACK FOUND IN BOILERS

The forms of corrosion previously considered can lead to the following results:

General Wastage. Overall reduction of metal thickness, especially in way of heating surfaces such as tube walls. Common in boilers having open feed systems and no protective treatment.

Pitting. This refers to irregular pits formed in the boiler metal by corrosive action and is by far the most serious form of corrosion normally found on the water side of marine boilers. It is often found in the boiler shell in way of the water level. Usually it is due to poor storage procedures when the boiler is shut down. In high pressure boilers it may also be found inside screen and generating tubes, and can also take place in superheater tubes when priming or carry over has occurred.

Corrosion Fatigue. Cases of this may be found in water tube boilers where irregular circulation through tubes in high temperature regions has induced alternating stresses to be set up in the tube material, resulting in the formation of series of fine cracks in the wall. Corrosive conditions aggravate the condition.

Stress Corrosion. If a material subjected to fairly heavy alternating stresses is placed in a corrosive environment, any resulting pitting can cause sufficient concentration of stresses to lead to the formation of fatigue cracks. The bare metal so exposed will then be subjected to further corrosive action, causing the process to continue. The result of this form of attack is referred to as grooving and may be found in flanged plates. For example, the end plates of Scotch boilers, and the ogee rings used in the construction of some types of vertical auxiliary boilers, often show evidence of this action. It generally results from undue stressing caused by poor steam raising procedures.

Caustic Cracking. This is a form of intercrystalline cracking caused by water with a high level of caustic alkalinity coming into contact with steel which has not been stress relieved. It is very rare in marine boilers but may occasionally occur in way of leaking riveted joints or bolted fittings.

This form of cracking differs from that of corrosion fatigue in that the cracks follow the grain boundaries of the metal, whereas the fatigue cracks pass across these boundaries.

Caustic Corrosion. This form of attack can take place at high pressures due to excessive concentrations of sodium hydroxide. It tends to occur in high temperature regions where, due to the rapid evaporation taking place, the sodium hydroxide forms local concentrations, nearly coming out of solution and forming a thin film close to the heating surface. This breaks down the magnitite layer and then reacts with the steel to produce a soluble compound, which then deposits on the surface in the form of a layer of loose porous oxide.

These local concentrations can in some cases be sufficient to cause a significant reduction in the measured level of alkalinity. However if the evaporation rate is then reduced, the hydroxide is released back into normal circulation, and the alkalinity is apparently restored. This phenomenon is referred to as caustic hideout.

Hydrogen Attack. If the magnitite layer is broken down due to some form of corrosive action, then at high temperatures hydrogen atoms can diffuse into the molecular structure of the steel until they meet and combine with carbon atoms to form methane. This is a very large molecule and causes stresses to be set up along the crystal boundaries of the steel eventually leading to either sharp sided pits, or else cracks to form in the tube walls.

High Temperature Corrosion. This form of attack can occur when loss of circulation causes the metal to overheat in a steam atmosphere. Another form of this direct corrosive action can take place externally, usually in way of the superheater tubes and supports, due to the hot exhaust gases reacting with the superheater materials.

FORMS OF CORROSIVE ATTACK FOUND IN THE FEED SYSTEMS

In addition to normal corrosive action, the following special forms of corrosion may occur in various parts of the system:

Graphitization. When unprotected cast iron is exposed to sea water, the metal corrodes away leaving behind a skeleton comprised mainly of graphite flakes. These have virtually no strength, and when the surface is removed collapse into carbon dust.

De-Zincification. This can occur when brass with a high zinc content is in contact with sea water. As the brass corrodes away, the copper component of the alloy is re-deposited back onto the surface, so giving in effect the removal of the zinc.

Exfoliation. This occasionally occurs in feed heaters with cupro-nickel tubes. The nickel is selectively oxidized from the alloy, resulting in layers of copper metal and nickel oxide.

Ammonia Corrosion. Ammonia formed by the decomposition of hydrazine can enter the feed system. It will then dissolve any cupric oxide formed on copper or copper alloy tubes. It will not however attack the copper metal, and so if no oxygen is present the process ceases. However any oxygen present can now attack the copper metal to form a thin film of cupric oxide impervious to further attack by the oxygen.

Thus neither ammonia nor oxygen alone will cause serious damage, but a combination of the two can lead to leaking tubes in air ejectors and other such units in low temperature parts of the feed system. Also the released copper can now pass over into the boiler where it can lead to corrosive pitting in the tubes.

SCALE FORMATION

Scale forms in boilers due to the presence of various salts which come out of solution and deposit because of the effects of temperature and ensity. In simple terms the process may be considered in the following manner. When a steam bubble forms on a heating surface, the evaporation of water involved causes a local concentration of solids, some of which do not re-dissolve when the bubble escapes, but remain to form small circles of crystals on the surface. Repeated formations of these build up the scale deposit, often forming in a series of layers of differing nature and composition. The rate of scale formation will be increased by the presence of corrosion products and (or) oil. The latter even in small amounts will not only increase the rate of scale formation but also its insulating effect.

DISSOLVED SOLIDS FOUND IN WATER

As rain falls through the air it absorbs carbon dioxide, becoming in effect a very dilute solution of carbonic acid and so easily dissolves any alkaline minerals with which it comes into contact. Two such minerals are the carbonates of calcium and magnesium, which dissolve in the water to form bi-carbonates of calcium and magnesium. Other compounds readily dissolved are sulphates, chlorides, and nitrates of calcium, magnesium, sodium and silica, but in general these are less common.

DISSOLVED SOLIDS IN FRESH WATER

Although filtered to remove suspended solids and organic matter, shore fresh water still contains dissolved solids and gases, giving in general two main classes of water. Hard water, containing mainly calcium and magnesium salts, tends to be alkaline in nature but when heated leads to scale formation. Soft water, containing mainly sodium salts, tends to be acidic in nature and so can cause corrosion but does not form scale.

If a choice is available soft water is to be preferred for use in the boiler, as it can easily be made slightly alkaline by the addition of small amounts of suitable alkali.

DISSOLVED SOLIDS IN SEA WATER

While sodium chloride is hardly present in fresh water, it forms the major constituent of the dissolved solids present in sea water. This is due to its high solubility, which continues under normal boiler conditions to very high densities. Other salts, although present in much smaller amounts, can lead to corrosion or to the formation of scale. The analysis of a typical sample of sea water is given below.

Calcium Bicarbonate	180 ppm
Magnesium Bicarbonate	150 ppm
Calcium Sulphate	1200 ppm
Magnesium Sulphate	1900 ppm
Magnesium Chloride	3200 ppm
Sodium Chloride	25600 ppm
Total	32230 ppm

Calcium bi-carbonate is slightly soluble in cold water, but when heated above 65°C it begins to decompose, carbon dioxide being driven off, while the remaining calcium carbonate, being insoluble in water, deposits as a soft white scale.

Magnesium bi-carbonate is soluble in cold water, but when heated above 90°C it decomposes, the carbon dioxide being driven off. The resulting magnesium carbonate then undergoes a further change, breaking down to form more carbon dioxide, and magnesium hydroxide. The latter deposits as a soft scale.

Calcium sulphate is the worst scale forming agent in the water, depositing as a thin, hard gray scale at temperatures above 140°C, or at densities above 96 000 ppm.

Magnesium chloride can break down under boiler conditions to form magnesium hydroxide which deposits as a soft scale, and hydrochloric acid which can set up an active corrosive action with the boiler metal. It will thus rapidly lower the pH of the boiler water in the event of sea water contamination.

Sodium chloride normally remains in solution forming most of the boiler water density. It can however come out of solution at very high densities above 225 000 ppm, when it then deposits as a soft incrustation.

Other scale forming salts may be present in small quantities, but can in general be neglected, although in high pressure boilers problems can arise due to silica. This is not however usually introduced by sea water contamination.

DANGER OF SCALE IN BOILERS

When a boiler is steaming the temperature of the hot gases are well above that at which steel becomes plastic, and for safe operation it is essential that a rapid flow of water is maintained to transfer the heat away from the metal as quickly as possible. Thus any substance whether it is scale, oil, sludge, or a film of steam, that impedes the flow of heat from the metal causing its temperature to be raised above its normal operational value, in effect reduces the factor of safety of the whole boiler. So dangerous is this aspect of scale formation that such considerations as loss of boiler efficiency or increased fuel consumption due to the reduced rate of heat transfer fade into insignificance.

PROPERTIES OF WATER

Here those properties important in feed treatment are considered.

COLOUR

Due mainly to dissolved organic matter. Of little direct importance, but allowance must be made for it when carrying out tests using standard colour comparisons.

HARDNESS

This refers to those dissolved solids in the water that can lead to the formation of scale. Hardness can be subdivided into:
Alkaline Hardness. Sometimes referred to as temporary hardness, it is due to the bi-carbonates of calcium and magnesium, which are slightly alkaline in nature. They rapidly decompose upon heating, to form carbon dioxide and the corresponding carbonates which then deposit as a soft scale, or sludge.
Non-Alkaline Hardness. Sometimes referred to as permanent hardness, it is due mainly to sulphates and chlorides of calcium and magnesium, which are acid in nature. They can deposit under certain boiler conditions to form scales of varying degrees of hardness.
Non-Hardness Salts. These consist mainly of sodium salts which remain in solution, and do not deposit under normal boiler conditions.

ALKALINITY

Solutions of which water forms a part, contain hydrogen ions and hydroxyl ions. When these are present in equal amounts the solution is said to be neutral. When there is an excess of hydrogen ions it is acid, and when an excess of hydroxyl ions

Solution	Grammes of H^+/litre of Solution	pH Value
Strong Acid	$1 = 10^0$	0
Neutral	$\frac{1}{10\,000\,000} = 10^{-7}$	7
Strong Alkali	$\frac{1}{10\,000\,000\,000\,000} = 10^{-14}$	14

Fig. 55 pH Value

it is alkaline. Keeping the water in a slightly alkaline condition reduces corrosion. The level of acidity or alkalinity is usually expressed in terms of its pH value. This is basically a measure of the hydrogen ion concentration in the solution; for convenience the very small values involved are expressed in terms of the logarithms of their reciprocals.

Thus, as shown in Fig. 55,

pH = Logarithm of the reciprocal of the hydrogen ion concentration in the solution

It should be noted that as the reciprocal is being used, the pH value increases as the actual hydrogen ion concentration decreases.

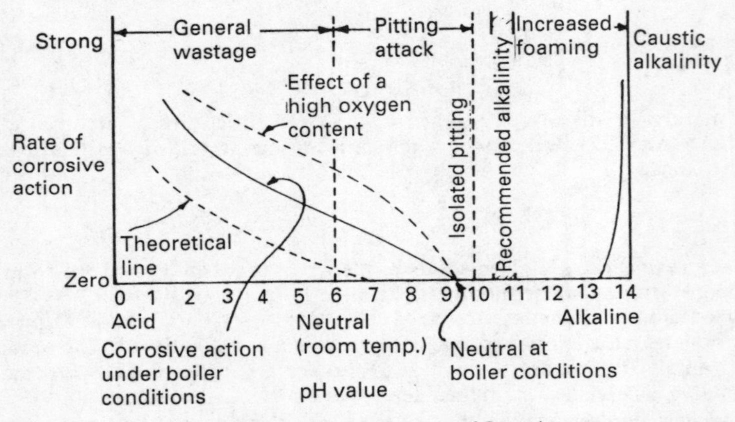

Fig. 56 Effect of pH Value upon Rate of Corrosion

The rate of corrosion is closely linked to the pH value of the solution, as indicated in Fig. 56, which shows how corrosive activity in a boiler varies with the pH value of the boiler water.

At a pH value of 7 theoretically no corrosion should occur, but in the boiler due to various factors the curve is lifted, so that a pH value of between 9 and 10 becomes the point where corrosion ceases. Thus by maintaining the pH value of the boiler water between 10·5 and 11 corrosion is kept to a minimum. Too high a pH value can lead to the water foaming in the stream drum, and also to possible caustic attack upon the boiler metal.

DISSOLVED SOLIDS

Various substances can dissolve in water so increasing its density. In marine practice the level of these dissolved solids is measured in the following ways:

TOTAL DISSOLVED SOLIDS

This refers to all the substances, harmful and otherwise, dissolved in the boiler water. A true measure of this density involves a complex process not used at sea; instead, for densitites above 2000 ppm, a hydrometer is used, while below this value an electrical conductivity meter is preferred.

CHLORIDES

Any sodium chloride will remain in solution in the boiler unchanged either by normal boiler conditions, or by chemical treatment. Therefore any increase in the chloride level will be an indication of contamination, either directly due to sea water leakage, or indirectly due to chlorides formed as the result of reactions between the treatment chemicals and the hardness salts.

Because of this and the fact that the chloride level can be measured with reasonable accuracy by a chemical test that can be used at sea, it is usually taken as the main guide to the level of density in high pressure boilers.

TREATMENT FOR MARINE FEED WATER

Ideally only distilled water should be used, all dissolved solids and gases being excluded from the feed system and boiler. Economically this is not practicable at sea and some form of treatment must be provided. This can be divided into external and internal forms of treatment.

External. This refers to treatment of the water before it enters the boiler; it can be mechanical, chemical, or both.

Mechanical. For pressures below 2000 kN/m^2 dissolved oxygen does not cause serious problems provided the boiler water is maintained in a slightly alkaline condition, thus open feed systems can be used. However in all cases it is advisable to avoid feeding cold water direct to the boiler as this introduces considerable amounts of dissolved oxygen. For pressures above 1750 kN/m^2 it is recommended that a deaerator is used to heat the feed water.

Only evaporated make-up feed should be used. However when, in the case of low pressure auxiliary boilers, raw make-up feed must be used it should be limited as far as possible to soft water. Under no account should river or lake water be used either to feed boilers or evaporators, due to the unknown industrial contaminants which may be present. Apart from being used as feed for evaporators, sea water should be excluded from the system at all times.

Filters can be used to remove suspended solids and oil from the feed water.

Chemical. This is referred to as water softening. The calcium and magnesium salts being precipitated, or converted to sodium salts by the addition of suitable chemicals before it enters the boilers. The most common water softening process is the Lime-Soda method of treatment, but it is of little use for marine practice as, to operate properly, special precipitation tanks are needed. The limited space of the engine room does not permit these to be fitted, while double bottom tanks are not suitable as the movement of the ship in a sea-way would not allow the precipitated sludge to settle out.

In some cases chemical treatment is added to water entering evaporators to ensure only soft scales are deposited.

With high pressure boilers there is increasing use of demineralization treatment for evaporated make-up feed before it enters the main feed system.

Prevention of Scale in Evaporators. This can be largely prevented by operating the evaporator at a sufficiently low temperature and density, due to the fact that calcium sulphate remains in solution at temperatures below 140°C and at densities below 96 000 ppm, and magnesium hydroxide will remain in solution at temperatures below 90°C. However calcium carbonate tends to form scale even below 80°C and it is usual to use chemical treatment to limit the formation of this deposit. Here the most commonly used forms of treatment are briefly described.

Ferric Chloride. This keeps the pH of the water at a value below that at which calcium carbonate, or magnesium hydroxide can decompose to form scale. Sodium bi-sulphate, or hydrochloric acid plus a corrosion inhibitor may be used as alternatives. A bonded rubber lining will protect the evaporator shell from the corrosive effects of the acidic water.

Sodium Polyphosphate. This consists of a mixture of various forms of sodium phosphate, and can be used to prevent scale formation in evaporators working at temperatures below 80°C. It is usual to add small amounts of sludge conditioning agents in the form of coagulants and anti-foams. This form of treatment is suitable for use in evaporators producing potable water.

Polyelectrolytes. These can be used to prevent the formation of scale in evaporators even when working at temperatures above 80°C. Again safe to use in the production of potable water.

REMOVAL OF SCALE FROM EVAPORATORS

Deposits of scale can be removed by mechanical cleaning or by thermal shock. Chemical cleaning can be used as an alternative to this, either sodium hydroxide or EDTA being used to remove calcium sulphate scale. Mild acids such as citric or sulphonic acids are used for the removal of calcium carbonate and magnesium hydroxide deposits.

DEMINERALIZATION TREATMENT

This refers to a process whereby ions in the water are exchanged for ions held in special resins. There are two basic forms of these exchange resins.

Cation Exchange Resins. Cations are ions with a positive electrical charge. These resins are charged with hydrogen ions by flushing through with a strong sulphuric or hydrochloric acid solution. The hydrogen ions can then be exchanged for other cations, those of calcium, magnesium, and sodium contained in the feed water.

Anion Exchange Resins. Anions are ions with a negative electrical charge. These resins are charged with hydroxyl ions by flushing through with a strong alkaline solution of sodium hydroxide. The hydroxyl ions can then be exchanged for other anions, those of chlorides, sulphates, and bi-carbonates contained in the feed water.

Mixed Bed Exchangers. These are usually referred to as demineralizers. They contain both cation and anion exchange resins, thus combining the two actions to remove, say, sodium chloride from the water. The cation resin will exchange hydrogen ions for the sodium ions, while the anion resin will exchange hydroxyl ions for the chloride ions. The released hydrogen and hydroxyl ions then combine to form water, which is now said to be demineralized.

Recharging. After a period of use the exchange resin becomes exhausted and needs to be recharged. In a mixed bed exchanger this can be done by back flushing to separate the different resins, these being of different size and with different relative densities. The separated resins are flushed through with the corresponding acid or alkaline solution.

A salinometer is fitted to the demineralizer outlet, and the exchange column recharged when this shows a density reading above 1 ppm.

INTERNAL FEED TREATMENT

The basic objectives of this treatment are as follows:

(1) To keep the boiler water in a slightly alkaline condition in order to reduce corrosion to a minimum.

(2) To precipitate any scale forming salts which may be in solution, and then keep them suspended in the boiler water.

FEED TREATMENT CHEMICALS

The following chemicals are in general use at sea for the treatment of boiler feed water.

Sodium Hydroxide. NaOH. The common name for this is caustic soda. It is a strong aggressive alkali which, in excess, can produce a condition known as caustic alkalinity in the boiler water.

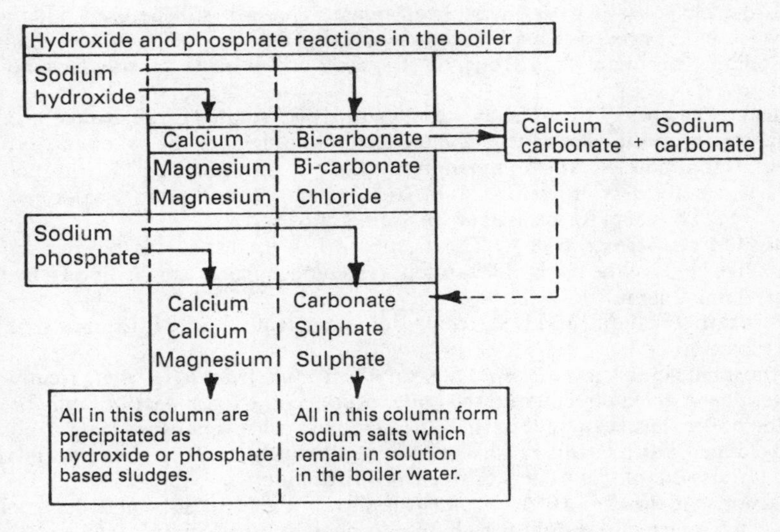

Fig. 57 Reactions of the Treatment Chemicals

Sodium hydroxide reacts with magnesium chloride so neutralizing this active corrosive agent, converting it into magnesium hydroxide, a harmless precipitate, and sodium chloride which remains in solution. Other reactions are indicated in Fig. 57. It should be noted that sodium hydroxide does not react directly with calcium sulphate.

This chemical is supplied either as a white solid, often in flake form which readily dissolves in water, evolving heat and producing a very strong alkaline solution, or as a concentrated solution which is diluted for use as required.

These solutions of sodium hydroxide must be handled with extreme care, gloves and goggles being worn when working with it.

Sodium hydroxide should be stored in cool dry conditions in tightly sealed containers as, upon exposure to the atmosphere, it absorbs carbon dioxide and converts to sodium carbonate. It is therefore not suitable for inclusion in standard mixtures, and is normally stored separately although used in conjunction with the other treatment chemicals.

Sodium Carbonate Na_2CO_3. This has the common name of soda ash. It is an alkaline substance and reacts with any non-alkaline hardness salts, either precipitating them, or converting them to their corresponding sodium salts which will remain in solution. Any excess sodium carbonate also remains in solution providing the alkalinity required to reduce corrosion.

At boiler pressures above 1400 kN/m^2 some of the sodium carbonate will decompose to form sodium hydroxide and carbon dioxide. The proportion that decomposes increasing as pressure increases.

Sodium carbonate is supplied in the form of a white powder which dissolves in water to form an alkaline solution. The powder is safe to handle provided it is dry. It should be stored in a cool place in sealed containers, and should not be left open as undue exposure to the atmosphere allows the absorption of carbon dioxide, changing it to sodium bi-carbonate. This can still be used but more powder will be needed so making control more difficult.

Sodium carbonate forms one of the basic constituents of standard boiler compound.

Sodium Phosphate. Phosphates react with the sulphates of calcium and magnesium, precipitating them as their corresponding phosphates, or converting them to sodium salts which remain in solution, as shown in Fig. 57. Phosphoric acid is a colourless crystalline solid from which various sodium salts can be prepared. The usual forms used at sea are as follows:

Sodium Metaphosphate. $NaPO_3$. This is supplied in the form of a powder, or as glass-like chips in the form of sodium hexametaphosphate, usually known by the trade name Calgon.

It is safe to handle, and is extremely soluble in water where it forms a slightly acid solution.

Phosphate in this form is especially suitable for use in systems where treatment cannot be injected directly into the boiler drum, as the final reactions only occur in the boiler thus keeping feed lines, heaters, and other items free from deposits and sludge. It also results in a lower level of alkalinity than the other phosphates and is thus suitable for boilers of riveted construction.

Disodium Phosphate. Na_2HPO_4. Usually supplied in the form of an anhydrous salt, it is often referred to simply as sodium phosphate. In the form of a white powder it is safe and easy to handle provided it is kept dry. It should be stored in a cool dry place. Disodium phosphate dissolves in water to form a neutral solution. Suitable for use where it can be injected directly into the boiler drum and, when some other chemical is used, to provide alkalinity. When added to water containing caustic alkalinity, it combines with the sodium hydroxide to form trisodium phosphate thus preventing the formation of excess levels of free sodium hydroxide.

Disodium phosphate forms one of the basic constituents of standard boiler compounds.

Trisodium Phosphate. Na_3PO_4. Again usually supplied as an anhydrous salt, it is safe and easy to handle provided it is kept dry. It should be stored in a cool dry place. Trisodium phosphate dissolves in water to form an alkaline solution. It may be noted here that when added to the water it decomposes to form sodium hydroxide and disodium phosphate. If some of the water is then evaporated, as the strength of the solution increases the density and caustic alkalinity also increase causing the sodium hydroxide and disodium phosphate to re-combine to form trisodium phosphate again. It is this process which forms the basis of what is known as co-ordinated phosphate treatment.

SLUDGE CONDITIONING AGENTS

These can be grouped as follows:

Coagulants. In some conditions the precipitated particles can adhere to the heating surfaces and form a soft scale. To reduce this possibility the particles of sludge can be physically conditioned by adding coagulants in the form of organic chemicals such as starch or tannin, or inorganic chemicals such as sodium aluminate. There is also increasing use of synthetic organic polymers of high molecular mass, often known as polyelectrolytes. This term covers a wide range of different types, the most widely used at sea for this purpose being sodium polyacrylate, carboxymethl cellulose, and some polyacrylamides.

Coagulants are colloidal substances which, by reason of their relatively large surface area and the electrical charge they possess, cause the fine particles of sludge floating in the water to collect about them. The larger particles thus formed are then less likely to adhere and will remain suspended in the boiler water. Some coagulants can also help to reduce foaming, and will tend to keep any oil present in an emulsified form.

It is important that the water is maintained in an alkaline condition otherwise the electrical charge of the coagulant is neutralized and it will lose its coagulating effect to a great extent, tending to allow soft deposits to form in the boiler.

Anti-Foams. Foaming may be caused by high density, high levels of alkalinity, or oil. It may be reduced by adding small amounts of anti-foams in the form of complex organic compounds of high molecular mass, usually in the form of polyoxides or polyamides. These reduce the surface tension of the water, thus causing the thin walls of the vapour bubbles to rupture more easily, and preventing them from accumulating to become a mass of foam.

Dispersing Agents. These may be added to prevent any solid precipitates from uniting to form sizable crystals. The chemicals used take the form of organic materials such as tannin, or sodium lignosulfonates.

OXYGEN SCAVENGERS

These are added to remove any remaining traces of oxygen in the boiler water after the bulk has been removed by mechanical deaeration. The two chemicals mainly used for this purpose are:

Sodium Sulphite. Na_2SO_3. Supplied as a white powder it dissolves in water and when injected into the boiler drum reacts with any dissolved oxygen in the boiler water to form sodium sulphate.

Hydrazine. N_2H_4. This is normally supplied as a 35% solution in water. It smells of ammonia and must be handled with care.

It is injected continuously into the feed system, reacting with dissolved oxygen to form nitrogen and water.

OTHER CHEMICALS USED

The following may be used in low pressure boilers:

Sodium Sulphate. Na_2SO_4. This may be added, with the initial starting up dose of boiler feed water chemicals, to boilers of riveted construction in order to reduce the possibility of caustic cracking.

Magnesium Sulphate. $MgSO_4$. Used as a sludge conditioner. Common name epsom salts.

FORMS OF INTERNAL CHEMICAL TREATMENT

The various treatment chemicals are combined in a number of ways to give the main forms of internal chemical treatment used at sea. The form and limits of treatment used are largely governed by the operating pressure of plant concerned. These pressures can be grouped as follows:

Low Pressure Tank Boilers. For boilers working at pressures of up to 1400 kN/m^2 a suitable treatment consists of sodium carbonate, which is used both to provide alkalinity and to precipitate hardness salts. Magnesium sulphate used as a sludge conditioner greatly helps this form of treatment by keeping the precipitated particles of sludge from depositing onto the heating surfaces.

However due to the decomposition of sodium carbonate that takes place at higher pressures its use as the sole means of treatment is limited to this low pressure range, otherwise the sodium hydroxide formed would lead to excessive levels of alkalinity.

Sodium sulphate may be added with the initial dose to boilers of riveted construction to reduce the risk of caustic alkalinity leading to caustic cracking in way of leaking rivets.

Medium Pressure Tank Boilers. For boilers working at pressures of up to 1750 kN/m^2 a very common form of treatment consists of a mixture of sodium carbonate, sodium phosphate, and a suitable sludge conditioner.

The sodium carbonate is now basically used to provide alkalinity and to deal effectively with any magnesium chloride which may be present in the water, while the phosphate is used to precipitate any hardness salts entering the boiler. It deals effectively with calcium sulphate.

For riveted boilers, or where the treatment cannot be injected directly into the boiler, sodium metaphosphate which gives the lowest level of alkalinity is generally used. In other cases disodium phosphate is preferred as it forms a neutral solution, so allowing the carbonate to control the level of alkalinity. In the presence of caustic alkalinity the disodium phosphate combines with the excess sodium hydroxide to form trisodium phosphate, thus preventing undue levels of caustic alkalinity forming in the boiler water.

Various sludge conditioning agents may be added, the most common being either starch or tannin.

Provided the boiler water is maintained in a slightly alkaline condition, at these low and medium pressures, dissolved oxygen does not present any serious corrosion problems, so even with open feed systems no effort is made to supply chemical de-oxygenation.

With boilers of riveted construction, sodium sulphate may be added with the initial dose of chemicals to reduce the risk of caustic cracking.

The chemicals are often supplied in the form of a standard boiler compound consisting of sodium carbonate, disodium phosphate, and starch, in the proportions of 3, 4, and 1 by mass respectively.

POLYELECTROLYTE TREATMENT FOR BOILERS

In low pressure plant due to the low quality feed often used, the use of phosphate treatment tends to produce large quantities of sludge. This is comprised both of precipitated hardness salts and the phosphate, the latter thus increasing the total amount of sludge formed. Excessive levels of sludge makes it difficult to ensure that it is all removed by blowing down, as some tends to settle out and form deposits on the heating surfaces.

To avoid this there is an increasing use of a polyelectrolyte type of treatment. In very simple terms these may be considered to act like super-coagulants. The scale crystals still form, but then adhere to the polyelectrolyte instead of to the boiler surfaces or to each other. The result is a free flowing sludge formed without the need of phosphate to first precipitate the hardness salts, as is the case with ordinary coagulants such as starch.

Another advantage claimed for this form of treatment is that, together with a suitable alkali to provide the necessary alkalinity to the water, it can be added in liquid form so being easier to dose. The sludge formed has greater mobility than a corresponding phosphate based sludge, and so is less likely to form deposits and is easier to blow out. The polyelectrolyte can disperse any deposits which have already formed, and have the ability to absorb traces of oil. Less elaborate test procedures are required; only those for alkalinity and density need be carried out. It is however vital that the correct level of alkalinity be maintained as if too low, the electrical charge possessed by the polyelectrolyte will be neutralized and its effectiveness greatly reduced. If the alkalinity is too high it can result in caustic alkalinity as there is now no phosphate to combine with any excess hydroxide and this puts limitations on the use of this form of treatment for high pressures.

The most commonly used polyelectrolytes in this form of treatment include sodium polyacrylate, carboxymethl cellulose, and some polyacrylamides.

TREATMENT FOR WATER TUBE BOILERS

In the case of highly rated water tube boilers, the treatment, while having the same basic aims as that used for tank boilers, must be operated within finer control limits. The dissolved solids in the system must be kept to a minimum and only good quality make-up feed must be used, this being supplied from an efficient evaporation system. The dissolved gases must also be kept to a minimum, thus a closed feed system should be used, except at the lowest pressures, with this type of boiler, while a deaerator, an advantage for medium pressures, becomes essential for pressures above 4200 kN/m^2. All evaporated make-up feed must be deaerated before entering the boiler as, although evaporation removes the dissolved solids, the dissolved gases can remain in solution.

In addition to the previously stated aims of providing alkalinity and keeping the boiler water at zero hardness by precipitating any hardness salts present, in high pressure boilers the treatment must also deal with any last traces of dissolved oxygen, and seek to establish and then maintain a protective film of magnitite on the internal surfaces of the boiler. Finally another essential requirement is to maintain a high degree of steam purity.

MEDIUM TO HIGH PRESSURE WATER TUBE BOILERS

For boilers working at pressures up to about 6000 kN/m^2 a combination of sodium carbonate, disodium phosphate, and a suitable sludge conditioning agent is again often used for the basic treatment, while sodium sulphite or hydrazine is used to absorb any traces of dissolved oxygen which may enter the boiler.

HIGH TO ULTRA HIGH PRESSURE WATER TUBE BOILERS

This refers to pressures ranging from 4200 kN/m^2 up to 8000 kN/m^2. The latter represents the upper limit for pressures in use at sea at the present time. Due to the greater proportion of sodium carbonate that decomposes at these higher pressures, sodium hydroxide is usually preferred as it gives better control over the level of alkalinity.

Disodium phosphate is again used to precipitate any hardness salts which may be present and to prevent excess levels of alkalinity.

In high temperature regions of high pressure boilers there is a tendency for excess sodium hydroxide to concentrate in localized areas close to the heating surface. It can then break down the magnitite layer and expose the boiler metal to corrosive attack. This can be avoided by using low pH forms of treatment such as co-ordinated phosphate treatment, in which careful control is kept over the sodium hydroxide to sodium phosphate ratio. This is to ensure that there is always sufficient excess disodium phosphate to combine with any undue concentrations of sodium hydroxide and form trisodium phosphate which, being more soluble than the hydroxide, remains in circulation. However this form of treatment works so close to the neutral pH value of the boiler water that it may not provide sufficient protection against corrosive attack in the event of sea water contamination. It is sometimes recommended that in the event of this contamination the treatment should revert to the standard form, which although it uses the same chemicals operates at higher pH values.

In some cases trisodium phosphate is used both to provide alkalinity and to precipitate any hardness salts which may be present. Use of this single chemical has the advantage of reducing the risk of overdosing with sodium hydroxide which could lead to caustic alkalinity. However it cannot cope effectively with sea water contamination as the disodium content is used up precipitating the hardness salts, so leaving an excess of sodium hydroxide, and is not to be recommended for use in marine boilers. Better results may be achieved by using a modified form of this treatment referred to as polyphosphate treatment which consists of a mixture of disodium and trisodium phosphates.

In all cases small amounts of suitable sludge conditioning agents will be added with the phosphate to keep any precipitated sludge in a free flowing condition, and also to reduce foaming with its attendant risk of increased carry over.

At these higher pressures removal of any dissolved oxygen is essential; the main forms of treatment are as follows:

TREATMENT FOR DISSOLVED GASES

Dissolved oxygen and carbon dioxide both lead to corrosion taking place in boilers and feed systems. The regenerative condenser and (or) a deaerator will remove the bulk of these dissolved gases and chemicals can then be used to neutralize any last remaining traces.

CHEMICAL DE-OXYGENATION

The two main forms of this in general use are:

SODIUM SULPHITE SO_2

This is suitable for pressures up to 4200 kN/m². The sodium sulphite reacts with oxygen to form sodium sulphate. It is injected directly into the boiler drum, or into the feed line just prior to the drum, so as to maintain a reserve of sodium sulphite in the boiler water ready to neutralize any dissolved oxygen which may enter.

AMINE TREATMENT

This refers to compounds of nitrogen and hydrogen, that is they have an ammonia base. Amines will combine with oxygen to form water and nitrogen, and so do not increase the density of the boiler water. Hydrazine is the most commonly used amine for this purpose.

HYDRAZINE N_2H_4

This is normally used for boilers working at pressures above 3200 kN/m². It is injected continuously into the feed system, from whence it passes into the boiler. A small reserve of hydrazine is maintained in the boiler water to deal with any dissolved oxygen which may enter. Some of the hydrazine evaporates and passes over with the steam thus providing protection to steam spaces and feed lines as well as to the boiler. It does this by absorbing oxygen and, by providing alkaline conditions in the feed system, gives protection against corrosion due to carbon dioxide.

Hydrazine is also beneficial in that it will react with ferric oxide and return it to the form of magnitite, which will then form a protective film on the boiler surfaces. It should be noted that this reaction can delay the establishment of the desired hydrazine reserve in the boiler when it is being put into service.

At temperatures above 350°C, which are usually only reached in the superheater, excess hydrazine can decompose to form ammonia and nitrogen. The ammonia passes over with the steam to go into solution with the condensate so producing alkaline feed conditions. This in general is beneficial but can, in the presence of dissolved oxygen, cause trouble with copper or copper alloys due to ammonia corrosion. This can occur in low temperature regions of the feed system, such as condenser air cooling sections and air ejectors if excessive amounts of hydrazine are used. The reason the oxygen can exist in a free state in these regions is due to the fact that hydrazine does not react with oxygen at temperatures below 50°C. In many cases in order to avoid this problem activated hydrazine is used, this contains a catalyst which enables the hydrazine to react with oxygen at temperatures below 50°C.

Hydrazine is normally injected into the discharge of either the extraction pump or the deaerator. The rate of injection is controlled to form, and then maintain, a small reserve of hydrazine in the boiler water.

Hydrazine is highly toxic in its concentrated form even by inhalation, and is thus normally supplied in the form of a 35% solution with water. This is colourless, smells of ammonia and is strongly alkaline. It can damage eyes and skin and so must be handled with care.

NEUTRALIZING AMINES

This term while including hydrazine is often used to refer to a group of very volatile amines which, if injected into the feed system, upon entering the boiler immediately evaporate. They then pass over with the steam to go into solution with the condensate, where they provide alkaline conditions and will chemically scavenge carbon dioxide by combining with it to form an amine bi-carbonate.

Of these volatile amines the most commonly used at sea are cyclohexylamine and morpholine. These remain stable to higher temperatures than does hydrazine and so reduce the possibility of ammonia attack on copper alloys in the feed system. They are injected with the hydrazine which must still be used, though in reduced amounts, to protect the boiler. This is because the volatile amines evaporate so quickly upon reaching the boiler that it is not possible to maintain a reserve of them in the boiler water.

Like hydrazine they must be handled with care. The rate of injection is usually controlled so as to maintain the pH of the feed system between 8·4 and 9·2. This value being measured by means of a pH meter.

FILMING AMINES

These may be used to protect auxiliary steam systems where the cost of injecting volatile amines would be too expensive. A filming amine such as octadecylamine, is injected continuously into the steam line where it leaves the boiler, so forming a protective film inside the steam pipes. This protects the metal from corrosion where wet steam conditions exist in the presence of carbon dioxide.

CHAPTER 10

Application and Control of Chemical Treatment

The treatment must always be added as a solution, the chemicals, where necessary, being added slowly to hot water at about 65°C in a suitable vessel stirring continuously until dissolved. The solution may be added via the top manhole when the boiler is open, or by suitable dosage equipment when closed or steaming. One form of this equipment consists of a dosing vessel which, filled with solution, can then be flushed through with water from the feed line so injecting the treatment into the steam drum. The other type consists of an electrically driven metering pump which pumps the dose into the drum. This type is also fitted for the injection of amine treatment.

If dosing equipment is not available the solution may be added to the hotwell of an open feed system, after the filter section. With a closed system it can be added via the auxiliary feed line, this being done by removing the cover of a suitable suction valve as close to the auxiliary feed pump as possible. Any high pressure feed filters must be by-passed.

After dosing, all equipment should be washed and pipe lines flushed through.

SAFETY PRECAUTIONS

The chemicals should be stored in a cool dry place, which can be secured against unauthorized entry. All containers must be clearly labelled as to their contents and only used for that chemical.

Preparation of the various doses should be carried out adjacent to the store room, in a well ventilated space having suitable supplies of both hot distilled water and cold fresh water, together with a sink and drain.

The chemical should always be added to the water, so forming a very dilute solution which is gradually strengthened. No naked lights must be allowed while mixing as flammable gases may be given off.

Personnel should be warned both verbally and by printed notices about the hazards of handling chemicals. Protective clothing, especially gloves and goggles, must be readily available.

In the event of a chemical burn, immediately place the affected part under cold running fresh water. Gently remove contaminated clothing and continue to flood the injured part for at least ten to fifteen minutes as measured by the clock. Any delay greatly reduces the effectiveness of this treatment.

INITIAL DOSAGE

Whenever the boiler is being filled, preferably with hot water, a suitable quantity of the appropriate chemicals should be added. Steam is then raised in the normal manner.

STEAMING DOSAGE

Subsequent dosing should be carried out as necessary to keep the test results within their recommended limits, any adjustments to the dosage rate being in accordance with instructions normally supplied. If none are available then increase or decrease the previous dosage rate by 10% until the limits are restored. Do not add large quantities of chemicals at one time; it is better to change the frequency of dosing than the amount. If scumming or blowing down is required it should be done before the chemicals are added.

CONTROL OF TREATMENT

This is done by a series of tests carried out with samples of boiler water. These should be taken from each boiler at least once daily, more frequent samples being taken in the event of contamination or when, due to unsatisfactory test results, the dosage rate has to be adjusted. To obtain a more representative sample from the boiler, clear the sampling line by allowing water to run to waste for a few minutes, then rinse out the sample bottle a number of times from the same line before final collection is made. Replace the stopper under the running water to ensure the bottle is completely filled and all air expelled. Do not use a glass bottle when a silica test is to be carried out with the sample. When using a salinometer pot on a low pressure auxiliary boiler a similar procedure should be followed, the pot being rinsed out a number of times with boiler water before the sample is finally collected. Testing should be carried out within half an hour of the sample being taken unless it can be kept in a completely filled and sealed bottle until required. The bottle should be clearly labelled with the date, time, and from which boiler or part of the system the sample has been taken.

When water at boiler pressure is drawn off into a salinometer pot at atmospheric pressure, it must reduce its heat energy to allow its saturation temperature to fall to 100°C so that it will correspond to the new pressure. It does this in a process known as flash off, by allowing some of its mass to flash off into steam, the latent heat required for this being obtained from the remaining water whose temperature is thus lowered. This process continues until the water reaches a temperature of 100°C. The steam flashing off leaves its dissolved solids behind in the water remaining in the pot thus distorting the results obtain from tests carried out with this sample of water. For pressures below 1400 kN/m^2 this effect may be neglected, but with higher pressures either a cooling coil must be used to reduce the temperature of the water to below 100°C before it is drawn off to atmospheric pressure, or alternatively a pressure type sampling vessel may be used. Here the water is drawn off under pressure and then allowed to cool before testing begins.

Samples should not be taken too soon after the boiler has been put into service, or after dosing with treatment chemicals. A period of at least one hour should be allowed for conditions to settle down.

All apparatus used for testing should be rinsed out with distilled or evaporated water and, if greasy, washed with a soapless detergent. The use of graduated droppers filled by means of rubber teats is to be recommended, and the use of pipettes filled by mouth should be avoided as many of the chemical reagents used in the tests are poisonous. Some test kits supply these reagents in tablet form, the quantity then being obtained by the number of tablets used instead of in ml. as measured by graduated burettes and droppers. Unless stated otherwise, filter the sample using a glass funnel and a suitable filter paper.

Any marked changes in the test results should be immediately investigated and if, as a result of these changes, the readings lie outside the recommended limits the appropriate action should be taken.

As the chemical treatment is designed to keep the boiler water slightly alkaline, at zero hardness, and for higher pressures with no dissolved oxygen, the tests carried out must be sensitive to changes in these areas. Contamination of the feed water, or blowing down, are the usual reasons for changes to occur in the test readings. However allowance must also be made for changes in boiler load, for recent chemical dosage, and for faulty test procedures. The last of these may be due to aged reagents or to contamination of the sample itself. It is thus advisable, before taking any direct action because of the readings, to obtain a fresh sample and carry out a new series of tests. This is especially the case when the observed changes do not fit the normal pattern, or if only a single test result has changed. See Table 1.

Table 1. Chart of Changes in Test Results

Test	Sea Water Contamination	Fresh Water Contamination	Leakage below Boiler Water Level
Hardness	Increase	Increase	Decrease
T.D.S.	Increase	Increase	Decrease
Chloride	Increase	Slight Increase	Decrease
Alkalinity	Decrease	Slight Increase	Decrease
Phosphate	Decrease	Decrease	Decrease

In addition to causing changes in the results of these tests, sea water contamination may cause a rise of the water lèvel in the reserve feed tank, not in keeping with normal operation, and an increase in the salinometer reading. It should be noted however that the latter can also be due to dirty electrodes, or to excess hydrazine in the system. Leakage below the water level in the boiler may be indicated by excessive make-up feed, by difficulties with combustion and, where leaking valves are involved, by hot blow down or drain lines.

The tests normally carried out daily with phosphate treated boilers are as follows:

HARDNESS

This measures the presence of any hardness salts in the sample. A soap test may be used but there is increasing tendency to use the more accurate EDTA test. A high reading points to contamination or to the phosphate reserve being too low, thus giving the possibility of scale formation.

ALKALINITY

This is to ensure that the boiler water is being maintained in a slightly alkaline condition to keep corrosion to a minimum. Because of the difficulty of measuring the pH value acccurately it is normal practice to use a phenolphthalein test which measures the amounts of hydroxides and carbonates in the sample. It should be noted that bi-carbonates cannot exist under boiler conditions, so none will be present in a sample of boiler water provided it is not exposed to the air too long before testing.

If the level of alkalinity is too low then corrosion can occur; if too high then foaming can take place in the drum, and if due to hydroxides the resulting caustic alkalinity can in some cases lead to caustic attack.

PHOSPHATE TEST

This is to measure the amount of phosphate in the sample, as a reserve of phosphate should be maintained in the boiler water ready to neutralize any hardnesss salts which may enter. A number of tests are available, one in common use being the vanado-molybdate test.

If this reserve is too low then it allows any hardness salts in the water to deposit as scale on the heating surfaces and it can also, under some conditions, lead to excessive amounts of hydroxide forming in the boiler water. Too high a reserve can lead to foaming in the drum, while overdosing with phosphate causes undue amounts of sludge which can settle out and form deposits on the heating surfaces.

CHLORIDES

This is to measure the amount of chloride in the sample. Since under normal boiler conditions these remain in solution and are also not affected by the treatment chemicals, any marked change in the chloride level can be an important indication of either sea water contamination, or of boiler leakage. A silver nitrate test is used. Too high a chloride level indicates that undue amounts of dissolved salts are present in the boiler water, leading to possible deposits and/or foaming which tends to increase the carry over of water droplets in the steam.

TOTAL DISSOLVED SOLIDS

This is a measure of all the dissolved solids in the water including the treatment chemicals, thus it is not such a good indicator of sea water contamination as the chloride test. Densities above 2000 ppm can be measured by means of a hydrometer, while for lower values an electrical conductivity meter is used.

Excessive density leads to foaming and or deposits.

SLUDGE

This refers to the precipitated particles formed as a result of the chemical treatment and which normally remain suspended in the water. It must not be confused with density which is due to solids still dissolved in the water. The sludge content cannot be measured directly by any of the chemical tests used at sea, but if allowed to accumulate can lead to foaming and eventually to deposits forming on the heating surfaces. Thus it is necessary that the boiler be blown down to reduce the sludge level at the recommended intervals.

SODIUM SULPHITE OR HYDRAZINE

When chemical de-oxygenation is being used the appropriate test will be necessary to ensure that an adequate reserve of the chemical is being maintained. If the result of the test is too low then corrosion can occur; if too high then, in the case of sulphite, an excessive density can arise, while with hydrazine an excessive amount of ammonia can be formed which may lead to attack upon copper alloys in the feed system.

LIMITS OF TREATMENT

The results of these tests must lie within certain recommended limits which can vary greatly from one type of boiler to another. However in general the higher the operating pressure of the boiler the smaller these limits will be. Typical values for boilers operating at various pressure ranges, together with their associated feed water, are shown in Table 2.

Table 2. Limits for Test Results

		Boiler Pressure (kN/m^2)					
		Shell	Water Tube				
Test	Limits (ppm)	<1750	<1750	1750–3200	3200–4200	4200–6000	6000–8500
LIMITS FOR BOILER WATER (ppm)							
Hardness	$CaCO_3$	5.0 max	5.0 max	5.0 max	1.0 max	1.0 max	1.0 max
P-Alk.	$CaCO_3$	150–300	150–300	150–300	100–150	50–100	50–80
Chlorides	$CaCO_3$	1000 max	300 max	150 max	100 max	50 max	30 max
T.D.S.	$CaCO_3$	7000 max	1000 max	1000 max	500 max	500 max	300 max
Phosphate	PO_4	30–70	30–70	30–70	30–50	30–50	20–30
Hydrazine	N_2H_4	—	—	—	0.1–1.0	0.1–1.0	0.1–1.0
Sulphite	SO_3	50–100	50–100	50–100	20–50	—	—
Silica	SiO_2	—	—	—	—	—	6.0 max
Iron	Fe	—	—	—	—	—	0.05 max
Copper	Cu	—	—	—	—	—	0.02 max
pH		10.5–11	10.5–11	10.5–11	10.5–11	10.5–11	10.3–11
LIMITS FOR FEED WATER (ppm)							
Chlorides	$CaCO_3$	10.0 max	5.0 max	1.0 max	1.0 max	1.0 max	1.0 max
Oxygen	O_2	—	—	0.06 max	0.03 max	0.015 max	0.01 max
Ammonia	NH_3	—	—	—	—	—	0.5 max
Iron	Fe	—	—	—	—	0.01 max	0.01 max
Copper	Cu	—	—	—	—	0.01 max	0.005 max
pH		—	—	—	8.5–9.2	8.5–9.2	8.5–9.2

Notes:
1. EDTA test for hardness quoted; for soap test result should be zero.
2. Coordinated phosphate treatment, boiler pH 9.8–10.3.

AUTOMATION OF WATER TREATMENT

For boilers operating at low to medium pressures, slug dosing of the chemical treatment coupled with intermittent sampling and testing of the boiler water is satisfactory. However it cannot maintain a constant level of protection, as values increase after dosing and subsequently fall until the next dose of chemicals is injected. The rate of change is monitored by testing samples of boiler water at regular intervals, so that as the lower limits recommended for protection are approached more chemicals can be injected to restore the values to their upper limits. Thus, under normal operating conditions, a reserve of treatment chemicals exists within the boiler to cope with any incoming contaminants.

Should for any reason the rate of contamination suddenly increase, this chemical reserve might be used up before the next routine sampling, and so damage could result. For boilers working at high pressure this problem increases, as the chemical treatment is operated within finer limits, giving a smaller reserve of chemicals available within the boiler to cope with any increased contamination.

Protection can be provided by the constant monitoring of the feed conditions, with as frequent testing of the boiler water as is practicable. However, better protection is given by continuous monitoring of both the feed and boiler water conditions, so that any undue change can be immediately detected and then rectified. This arrangement can also be used in conjunction with the continuous injection of the treatment chemicals, which, coupled with an automatically controlled rate of blow down, allows a constant level of protection to be maintained for both the feed and boiler water.

While for slug dosing continuous monitoring of the feed water by a salinometer and a pH meter suffices, for automatic control a more sophisticated monitoring system is required for both feed and boiler water, as indicated in Fig. 59.

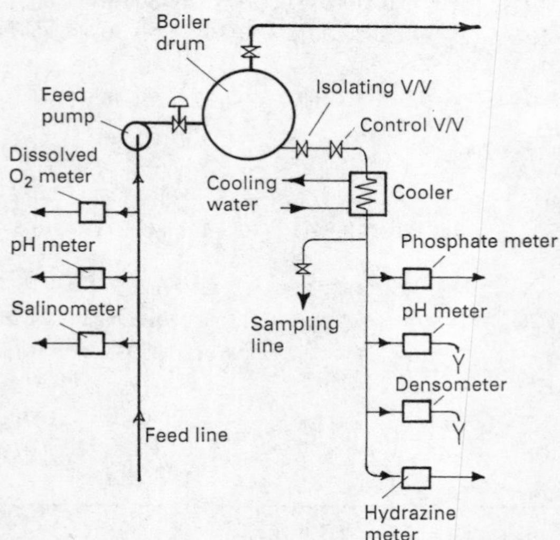

Fig. 59. Minimum Requirements for Continuous Monitoring of Feed and Boiler Water

The readings obtained can be displayed in the control room, so allowing constant observation of the boiler conditions, and if desired can be used to provide signals for the automatic control of the rates of chemical injection and of boiler blow down.

This form of control can also be used, especially in the case of low pressure boilers, to reduce manning levels.

Test Procedures

HARDNESS

Wanklyn Soap Test. A standard soap solution is added to a glass stoppered bottle containing 100 ml of filtered water. The bottle should be shaken vigorously after each addition of 0·2 ml of soap solution until a lather which persists for at least five minutes when the bottle is laid on its side is obtained. The soap solution reacts with any hardness salts present to form an insoluble scum, and not until all these salts have been precipitated can a lather form.

1 ml of soap solution will precipitate 10 ppm of $CaCO_3$

Thus ml of soap solution x 10 = Hardness in terms of ppm $CaCO_3$

EDTA Test. Pour 100 ml of filtered water into a porcelain dish. Add approximately 2 ml of ammonia buffer solution and 1 ml of Mordant Black in-indicator and stir. If hardness salts are present the solution will turn wine-red. Immediately add drops of EDTA solution stirring continuously until the red colour just disappears; the solution will now be blue.

1 ml of 0·02N EDTA solution will precipitate 10 ppm of $CaCO_3$

Thus ml of 0·02N EDTA solution used x 10 = Hardness in terms of ppm $CaCO_3$

ALKALINITY

Phenolphthalein Test. Pour 100 ml of filtered water into a porcelain dish. Add 1 ml of phenolphthalein solution and if the sample has a pH value greater than 8·4 it will turn pink. Add drops of sulphuric acid stirring continuously until the pink colour just disappears.

1 ml of 0·02N sulphuric acid will precipitate 10 ppm of $CaCO_3$

Thus ml of 0·02N sulphuric acid used x 10 = Alkalinity to phenolphthalein in terms of ppm $CaCO_3$

Methyl-orange Test for Total Alkalinity. Bi-carbonates do not show up in the phenolphthalein test, as they have a pH value less than 8·4 and so do not cause phenolphthalein to turn pink, thus if their presence is suspected in a sample of raw feed, carry out the following test using Methyl-orange.

Take the sample used for the phenolphtyalein test and to it add 1 ml of methyl-orange. If the sample immediately turns red it shows no bi-carbonates are present, if however it turns yellow it indicates the presence of bi-carbonates and drops of sulphuric acid are added until the solution turns red.

1 ml of 0·02N sulphuric acid will precipitate 10 ppm of $CaCO_3$

Thus total ml of 0·02N sulphuric acid used in both the phenolphthalein and methyl-orange tests x 10 = Total alkalinity in terms of ppm $CaCO_3$

Barium Chloride Test for Caustic Alkalinity. Pour 100 ml of filtered water into a porcelain dish and add 10 ml of barium chloride solution stirring well for two minutes. This precipitates any carbonates present in the sample, thus any pink colouration taking place when 1 ml of phenolphthalein indicator is added to the solution must be due to hydroxides only. Add drops of sulphuric acid until this pink colour just disappears. Disregard any reappearance of this pink colour.

1 ml of 0·02N sulphuric acid will precipitate 10 ppm of $CaCO_3$

Thus ml of 0·02N sulphuric acid used x 10 = Caustic alkalinity in terms of ppm $CaCO_3$

pH Test. Pour 100 ml of unfiltered water from a previously sealed bottle into two special 50 ml glass stoppered test tubes, one of which contains 0·2 ml of a suitable pH indicator. Immediately replace the stoppers and compare the resulting colouration against standard colour discs. The tube without the colour indicator being placed beneath the disc in use to offset the effect of any discolouration of the boiler water itself upon the observed result.

Various indicators may be used depending upon the pH range to be measured. Phenol red for pH values 6·8 to 8·4; Thymol blue for pH values 8·0 to 9·6; Thymol phthalein for pH values 9·3 to 10·6; Nitro yellow for pH values 10·3 to 11·6.

Great care must be taken to avoid the sample coming into contact with the air during the test, otherwise some of the carbonates may form bi-carbonates so changing the alkalinity level of the sample. Due to the difficulty of obtaining an accurate result from this test, it is generally recommended that where pH measurement is necessary an electrical pH meter should be used.

CHLORIDES

Silver Nitrate Test. The sample must be acidic and it is thus convenient to use the sample previously used in the phenolphthalein test, making it positively acidic by adding 4 ml of sulphuric acid. This is followed by 0·5 ml of potassium chromate indicator which turns the sample yellow if any chlorides are present. If this is the case add drops of silver nitrate solution, stirring continuously until the colour just changes to a reddish brown.

If the sample from the phenolphthalein test is not available, or if sulphite is present the sample should be prepared as follows:

Pour 100 ml of filtered water into a porcelain dish and add 1 g of potassium persulphate stirring until dissolved. This is followed by 1 ml of phenolphthalein indicator with sulphuric acid being added until the pink colour just disappears. The previously described silver nitrate test is now carried out.

1 ml of 0·02N silver nitrate solution will precipitate 10 ppm of $CaCO_3$

Thus ml of 0·02N silver nitrate solution used x 10 = Chloride content in terms of ppm $CaCO_3$

TOTAL DISSOLVED SOLIDS

By use of Hydrometer for TDS values above 2000 ppm. Pour about one litre of filtered water into a hydrometer jar. Add six drops of neutral alkyl sulphate as a wetting agent and stir gently. Hold the hydrometer by the top of its stem only and lower it gently into the sample. Any air bubbles clinging to the hydrometer can generally be removed by spinning it gently in the water. Read off the density from the hydrometer stem at the underside of the meniscus, making sure the instrument is floating freely and not touching the side of the jar. Take the temperature of the sample at the time of the reading, and if it does not correspond to the temperature given for the hydrometer, the density reading must be corrected by the appropriate formula or conversion table.

Great care must be taken to ensure that the hydrometer and jar are kept clean and free from grease. Lack of cleanliness is indicated by distortion of the meniscus and the apparatus should be cleaned.

Electrical Conductivity used for TDS values below 2000 ppm. Pure distilled water does not conduct electricity, but impure water does, its electrical conductance being proportional to the quantity and nature of the substances dissolved in it. Thus knowing the analysis of these substances the density can be measured in terms of electrical conductivity. However the salts normally found in sea water all have approximately the same level of conductivity and marine salinometers can be calibrated to read directly in terms of the ppm of dissolved salts present in the sample. Electrical conductivity varies with temperature thus the instrument must be provided with a temperature compensating device.

The salinometer may be a permanent unit giving a continuous reading, with probes mounted in the feed line to measure the feed density and, in some cases, in a special blow down line which enables the density of the boiler water itself to be measured. Alternatively, small portable salinometers may be supplied, these being used as follows:

Pour 250 ml of filtered water at a temperature of 20°C into a porcelain dish and add four drops of phenolphthalein indicator. Drops of acetic acid are now added stirring continuously until the pink colour disappears. Use this neutralized sample to wash out and fill the conductivity cell. Check that the salinometer is set for both correct density range and temperature. Press the operating button and zero the instrument. The density is then read from the scale either directly in terms of ppm of dissolved salts, or as electrical conductivity which can be converted to density by a constant;

Conductivity in siemen/metre x 70 = Dissolved salts in ppm.

When the water is alkaline or contains hydrazine it causes different levels of conductivity to be produced as compared to those of the usual salts thus leading to inaccurate readings. The sample should therefore be neutralized, or for fixed installations the water entering the sampling cell first passed through a demineralizing cartridge. This converts all the dissolved salts, together with the hydrazine, to hydrochloric acid, the density then being proportional to the conductivity due to this acid.

PHOSPHATE RESERVE

There are a number of tests available for this, two are in common use:

Vanado-molybdate Test. Pour 25 ml of filtered water into a 50 ml stoppered graduated cylinder and add an equal volume of vanado-molybdate solution. Stopper the cylinder and shake to mix the contents. Use this solution to wash out and then fill a 25 ml comparitor tube and fit the stopper. A similar tube is prepared as a blank with equal volumes of distilled water and reagent. Allow the colour to develop for at least three minutes, then using a colour comparitor match the sample tube against a standard colour disc with the blank placed beneath it.

If the initial sample is discoloured, add 2 g potassium persulphate to 50 ml of the water and boil until colourless. Cool and make up to 50 ml with distilled water. It is now ready for testing.

For auxiliary boilers two standard colour discs may be used; if sample colour is lighter than a 30 ppm PO_4 disc reserve is too low, while darker than an 80 ppm PO_4 disc indicates it to be too high.

Potassium Nitrate Test. Add 4 g of potassium nitrate to 50 ml of hot filtered water and shake until dissolved. Filter until a clear solution is obtained. 25 ml of this solution is heated to 40°C and 5 ml of ammonium molybdate added, the solution shaken and allowed to stand until the onset of cloudiness. Note the time taken for this, as the greater the phosphate content the quicker this reaction occurs; less than two minutes indicates a reserve of more than 80 ppm PO_4; more than five minutes indicates it to be less than 20 ppm PO_4.

HYDRAZINE RESERVE

Dimethylaminobenzaldehyde Test. In the presence of hydrazine this reagent produces a yellow colour. The intensity of the colour depending upon the amount of hydrazine present in the sample.

Any hydrazine in the sample can react with oxygen, thus every precaution must be taken to prevent undue contact with air during the test. The water should be cooled to about 25°C in a stainless steel cooler fitted with an extension tube. Collect a representative sample by inserting this tube into a 250 ml stoppered bottle, which is filled slowly from the bottom, withdrawing the tube and inserting the stopper under running water to avoid trapping any air bubbles. Immediately add 40 ml of this sample to one of two 50 ml stoppered colour comparitor tubes containing 10 ml of acidic 4-dimethylaminobenzaldehyde. If necessary the water should be poured quickly through filter paper. The other comparitor tube is prepared as a blank by adding 40 ml of distilled water to 10 ml of reagent. Allow ten minutes for the colour to develop and then, using a colour comparitor, match the sample tube against a standard colour disc with the blank placed beneath it. The value of the hydrazine reserve is given on the disc showing the same colour as the sample.

This test is limited to a maximum value of 0·25 ppm of N_2H_4; for higher values such as these used for boiler storage, the sample is diluted and the disc reading corrected to suit.

SULPHITE RESERVE

Potassium Iodate-iodide Test. As sulphite reacts with oxygen a similar procedure must be followed as for the hydrazine test when obtaining the sample.

Immediately pour 100 ml of this unfiltered water into a porcelain dish containing 4 ml of sulphuric acid and 1 ml of starch indicator solution. Drops of potassium iodate-iodide are added to the solution, stirring continuously. This reagent oxidizes the sulphite to sulphate and when the reaction is complete, any more reagent added to the solution causes the starch indicator to produce a blue colour. Immediately this occurs stop adding the reagent and note the amount used. The sulphite reserve is then calculated by means of the following formula:

$$\frac{\text{ml of iodate-iodide solution}}{\text{ml of sample}} \times 806 = \text{Reserve in ppm of } SO_3$$

AMMONIA IN THE FEED WATER

Nessler Reagent Test. This test is only necessary where hydrazine is being used to scavenge oxygen. The reagent is prepared from potassium iodide, mercuric chloride, and sodium hydroxide, by a process so involved that it is supplied already mixed and ready for use. When added to a sample containing ammonia it turns reddish brown, the intensity being proportional to the ammonia concentration. It should be noted that this reagent is highly poisonous.

Draw off a sample of condensate and pour it through filter paper to fill two 50 ml stoppered colour comparitor tubes. Add 2 ml of the Nessler reagent to one of these tubes and allow ten minutes for the colour to develop. The using a colour comparitor match the colour produced against the appropriate disc, with the other tube placed beneath it to offset any discolouration of the initial sample. The amount of ammonia is either read off directly from the disc, or calculated from the formula:

$$\frac{\text{Disc reading in ug}}{50} = \text{ppm of } NH_3$$

DISSOLVED OXYGEN IN THE FEED WATER

Indigo-carmine Test. This reagent produces a series of colours ranging from orange through pink to blue with increasing oxygen content. To ensure an accurate result great care must be taken throughout the test to avoid any air contamination of the sample.

The reagent must be prepared fresh for the test by adding 2 ml of potassium hydroxide solution to 8 ml of a solution of indigo-carmine reduced with glucose. Stopper the bottle and mix well. Keep undisturbed in the dark at 15° to 30°C for about 30 minutes until the colour has changed to lemon yellow. This is referred to as the leuco reagent and is only stable for about 12 to 15 hours.

Special stoppered colour comparitor tubes having a small inner tube, and known as modified Nessler cylinders are required for this test. The inner tube is completely filled with the leuco reagent and then sealed with a small glass ball positioned with the aid of a short length of tubing. Make certain no air bubbles remain.

The sample is cooled as close to the ambient air temperature as possible by passing it through a stainless steel cooler with an outlet pipe carefully arranged to prevent air bubbles from being retained. With the water running, place the Nessler cylinder over this tube so that it fills from the bottom and allow the water to run for at least two minutes. The cylinder is then slowly lowered until just clear of the sampling tube, and with the water still running, immediately slide a clean, freshly wetting glass stopper across and into the top of the cylinder, using a twisting motion to exclude any air bubbles. Holding the stopper firmly in place invert the cylinder so that the glass ball falls off and allows the reagent to mix with the sample. Invert a number of times to ensure all the reagent is thoroughly mixed and leave for five minutes for the colour to develop. The using a colour comparitor match the colour against the appropriate disc, this having a similar tube containing sample water only placed beneath it to offset any discolouration of the initial sample. The result is read off from the disc showing the same colour as in the sample tube. If given in ml/litre it can be converted to ppm as follows:

$$\text{ml/litre} \times 1 \cdot 5 = \text{ppm of dissolved oxygen}$$

TESTS FOR SILICA, IRON, AND COPPER

These tests are not usually carried out on board ship; instead samples of water are sent ashore for analysis. In some cases, in order to obtain more accurate information regarding the iron and copper content of the feed water, a stated volume of this water is passed through a special millipore filter cartridge. This is then sent ashore so that any particles filtered out can be analyzed.

EXPRESSION OF RESULTS IN TERMS OF CALCIUM CARBONATE $CaCO_3$

As the results from the various tests are ratios rather than absolute values, they must be related to some basic substance. For hardness, alkalinity, and chlorides where a number of substances are involved it is usual to express the results in terms of calcium carbonate, this being chosen because its molecular mass of 100 make for simplicity of calculation and of reagent strengths.

RECORDS

All test results should be recorded, together with times and amounts of chemical dosage, blow down, and addition of make-up feed. It is also advisable to note steaming condition, changes of load, soot blowing, and other similar activities, so that these can be correlated with any changes in the test results.

REMOVAL OF OIL FROM BOILER INTERNALS

If oil enters a steaming boiler, being detected either in the gauge glasses, or by known contamination of the feed system, immediately reduce the boiler load as much as possible. Where a scum valve is fitted the boiler should be scummed frequently, and in the absence of specific instructions for the boiler, increase the chemical dosage rate to keep the alkalinity and phosphate reserves near their recommended maximum levels. The alkali acts as a detergent causing the oil to form an emulsion and the phosphate and sludge conditioners then serve to keep these emulsified particles suspended in the water. If available, additional sludge conditioners should be added, especially those to reduce foaming. The use of magnesium sulphate may also prove to be beneficial. When no scum valve is fitted, use frequent flash-blowing of say five to 15 seconds duration, to remove the oily sludge formed by the chemical treatment.

Trace the source of the oil contamination and rectify immediately. If the contamination is heavy, shut the boiler down and subject it to an alkali boil-out.

It should be noted that if the contamination proves to be very heavy and (or) due to heavy fuel oil, specialist treatment involving either a vapour solvent process or acid cleaning, will be necessary in addition to an alkali boil-out.

PROCEDURE FOR AN ALKALI BOIL-OUT

Shut the boiler down and allow pressure to fall to about 350 kN/m^2; the scum valve can then be used to scum the boiler. Refill to normal water level and allow the boiler to cool down. Add water as necessary to keep the water level within the limits of the water level indicators. The air vent can be opened and, when the pressure is off the boiler, the top manhole doors can be opened, the normal safety precautions being followed. The boiler can then be run down, hosing down the surfaces as the water level falls. An alternative procedure, if no scum valve is fitted, is, after having removed the top manhole doors, to add feed water until it overflows through the top manholes. This will allow the bulk of the oil floating on the water surfaces to spill out, instead of remaining on the boiler surfaces as the water level drops.

Do not blow an oil contaminated boiler right down, otherwise as the water level falls the oily sludge will deposit, and then bake onto the still hot boiler surfaces.

When the oil contamination is only discovered after opening up, remove as much of the greasy deposits as possible from the internal surfaces by means of a high pressure water jet. Evidence of even small greasy deposits in the drums should always be treated with caution as more serious deposits may exist in the tubes, left behind by the falling water level when the boiler was emptied. Small areas can be cleaned manually, but do not use a grease solvent unless it is non-flammable and does not give off toxic fumes.

Having removed the bulk of the greasy deposits add a strong solution of trisodium phosphate, consisting of approximately 1 kg of chemical for every 2000 kg of water held by the boiler. Alternatively magnesium sulphate, or if preferred a proprietary boiling out compound, may be used. The chosen chemical dissolved in hot water is poured in through the top manhole door, which is then closed and the boiler filled in the normal manner until the water level indicators show three quarters full.

Steam is then raised to a pressure of about 350 kN/m^2, at which point the safety valve is eased so as to produce rapid circulation within the boiler. If a superheater is fitted use the safety valve on the outlet header for this purpose. When the water level shown in the glass has fallen by about 100 mm, close the safety valve and extinguish the fires. The scum valve should then be used to remove oily scum from the water surface. If no scum valve is fitted then flash blowing should be carried out.

After scumming, the boiler should be refilled to three quarter glass, and the burners flashed up to return the boiler pressure to 350 kN/m^2. The scumming procedure is repeated every two hours for the next 12 hours. The fires are then extinguished and the boiler pressure allowed to fall to 200 kN/m^2, when the boiler is finally scummed and then emptied by blowing down. When the pressure is off, the air vent is opened and the boiler allowed to cool down. It is then opened up and an internal examination carried out. If any traces of oil remain the procedure should be repeated.

When the boiler is clean, flush it out with fresh water then box up, refill and raise steam in the usual way. It should then be shut down and scummed and blown down until empty. The boiler is now ready to be refilled and returned to service in the normal manner.

CHAPTER 11

Cleaning and Storage of Boilers

If a shut down boiler is left partially filled with water severe corrosion can take place, often resulting in pitting along the water level and the top of the steam space. Thus when a boiler is to be shut down for a period of time, one of two basic procedures should be followed, the choice largely depending upon the expected length of time for which storage will be necessary.

WET STORAGE

The boiler should be completely filled and kept pressed up with alkaline water. This is suitable for storage periods of up to about three months, or when the boiler is to be held available for standby service and therefore must be ready for rapid return to operation.

DRY STORAGE

The boiler is completely drained and kept dry. This is suitable for storage periods of over three months, and where allowance can be made for time to prepare the boiler for return to service.

PROCEDURE FOR WET STORAGE

When shutting the boiler down for a few days, it will be convenient to fill it completely with hot alkaline water. The alkalinity being obtained by adding suitable chemicals a short time before the boiler is shut down. If the boiler density is high, or when storing for a longer period, the density should first be reduced by scumming and blowing down in the normal manner. The water level should then be raised as high in the glass as is consistent for safe operation, and raised to its boiling point to ensure that the chemicals thoroughly mix and complete their reactions. The boiler is then shut down and when the pressure is nearly off, the air vent opened and the boiler completely filled. With all air removed the air vent can be closed, pumping being continued until a slight hydraulic pressure is obtained, or alternatively when the arrangement allows it, the boiler can be flooded from the deaerator. This should provide a head of water sufficient to preclude the ingress of air into the boiler. When a superheater is fitted it must also be flooded. This can be done from the boiler, the superheater circulating valve being left open so allowing the water to rise some way up the circulating pipe, removing air and providing the necessary head.

A positive pressure must be maintained throughout the storage period, being topped up as necessary, or maintained by a suitable header tank, such as a deaerator.

In many cases a simmering coil is fitted in the boiler and this, supplied with steam from another source, enables the water in the boiler to be kept hot. This not only helps to maintain a positive pressure, but also keeps the gas side warm and dry.

All valves on the boiler, not actually required, must be firmly closed and the glands nipped up. Any steam lines and pumps not in use should preferably be drained, any exposed rods or spindles being protected from corrosion.

The alkalinity of water in the stored boiler should be checked at least once a week and additional chemicals added if necessary to restore it to the desired value.

Alkalinity can be provided in the stored boiler by supplying additional amounts of the chemical normally used to scavenge oxygen in the steaming boiler. This will consist of either sodium sulphite or hydrazine. The chemical in solution is injected either into the feed system or directly into the boiler drum, a short time before steaming ceases and as the boiler is being filled preferably with hot, deaerated water. When using hydrazine, it is essential to stop injection about 15 minutes before stopping the feed pump, otherwise a hydrazine rich mixture could remain in the feed lines and pump casing and cause damage to non-ferrous fittings.

Make sure that the boiler stored in a wet condition is never subjected to freezing conditions.

Before returning the boiler to service it should be completely drained and then refilled to normal level for flashing-up.

PROCEDURE FOR DRY STORAGE

Shut down in the normal way and allow the boiler to cool. It should then be completely drained. It is often advisable to open the steam drum doors and to hose the surfaces down with fresh water as the water level falls, especially if slight oil contamination is suspected. When the boiler is empty remove all manhole and handhole doors.

Inspect boiler internally noting its condition, and ensuring that any remaining traces of water are removed, then dry out the boiler by gentle heat, using hot air blowers or electric heaters. If necessary the boiler should be cleaned internally and any work required carried out. The heaters should be used when work is not in progress to prevent condensation forming in the boiler.

When all work is completed, fit new gaskets to all handhole and manhole doors, then place shallow trays of a suitable dessicant, such as silica gel, activated alumina, or calcium oxide, in drums and headers. Box up, replacing the doors, and making sure all valves are firmly closed. It is important to make sure that no water or steam can enter the boiler through leaking valves. If this cannot be prevented, then wet storage should be used instead.

Provided the boiler is completely sealed, the action of the dessicant in absorbing moisture eventually produces a slight vacuum in the boiler, which can be measured by a suitable U-tube arrangement, and providing a useful check on the internal condition. About every two months, or more frequently if any problems are suspected, open up the boiler and ventilate it. When safe to enter, check for any signs of water and renew the trays of dessicant, then box up again. The dessicant removed should be dried out by heating and then stored in a sealed container ready for use. Only a few kilogrammes will normally be required.

When access to headers is not convenient, or welded handhole doors are fitted, having completely drained the boiler, it should be dried out internally by the following procedure: Heaters are installed in the steam and water drums, the doors in the latter being closed by suitable air tight covers. The internal temperature of the boiler is then raised by the heaters until any remaining moisture has been evaporated and dispersed through the open steam drum doors, air vents and superheater circulating valve. The heaters are removed and shallow trays, of a suitable dessicant, placed in the steam and water drums, and boiler boxed up.

Another method of dry storage gaining increasing favour, is to fill the stored boiler with nitrogen under a slight positive pressure. The gas is stored in bottles and delivered to the boiler through a pressure reducing valve which enables a constant gas pressure to be maintained in the boiler during the storage period. Shut the boiler down in the normal manner and allow to cool. When the boiler pressure has fallen below the gas supply pressure, the nitrogen is admitted through a suitable connection to the boiler, where it displaces the boiler water which drains out through the run down valves. When all the water has been displaced, the superheater and economiser drains should be cracked open and the nitrogen allowed to vent for a short period. When admitted to the boiler the dry nitrogen will absorb any remaining moisture and this purging will help to dry out the boiler. Provided the boiler valves are firmly closed, the loss of nitrogen during the storage period should be slight and easily made up from the gas storage bottles.

To return the boiler to service, shut off and disconnect the nitrogen supply. The air vents and superheater circulating valves are then opened and water supplied to the boiler to bring the water level to that required for flashing up. The normal venting that takes place as steam is raised is sufficient to displace any nitrogen left in the boiler.

PROCEDURE FOR GAS SIDE STORAGE

These must be thoroughly cleaned, by water washing if necessary. The furnace and gas passages should then be inspected to make sure all deposits have been removed. They should then be dried out with gentle heat using hot air from steam air heaters or suitable blowers. As an alternative electric heaters can be used. Then close all air checks and dampers. The funnel top should be closed with a suitable cover. To prevent condensation the furnace heaters should remain in use during the storage period.

CHEMICAL CLEANING OF INTERNAL BOILER SURFACES

This is usually carried out when commissioning new boilers and when deposits formed on the internal surfaces cannot effectively be removed by mechanical cleaning.

Pre-commission cleaning consists of an alkali boil out, to remove traces of oil and dirt, followed by the main acid wash carried out at a temperature of about 80°C to remove rust and mill scale. It is recommended that a 3% to 5% solution of citric acid together with a suitable inhibitor is used as, although more expensive, it is less aggressive than other acids used and so reduces the likelihood of serious attack on the boiler metal during the cleaning process. The temperature is obtained by mixing the acid with hot water, the resulting solution being circulated through the boiler for some hours. During this time parts of the boiler not heating up can indicate poor circulation, and restrictions may have to be fitted in tubes to ensure a more even distribtuion. If regular tests show an undue increase in the ferrous content on samples drawn off from the boiler, stop the process immediately.

With acid cleaning complete, the boiler is drained and then flushed out with a dilute solution of phosphoric acid. This is known as a chelating rinse and serves to prevent surface rusting. Using warm water of as good a quality as possible, continue to flush out the boiler until a neutral solution is obtained. An internal inspection should then be carried out, and a hose used to flush out any parts where poor circulation was suspected. The boiler is refilled and stored under pressure with water containing hydrazine.

Except for the alkali boil out, a similar procedure is employed for the feed system.

Removal of deposits from boilers which have been in service requires a more aggressive acid, and usually a 3% to 5% solution of hydrochloric or sulphuric acid is used. A suitable inhibitor forming a protective film on the metal surface must be included in the solution. All internals must be removed, and any sections of the plant not to be cleaned securely blanked off. Acid cleaning then proceeds as described for new boilers. Samples drawn off are tested, and when the level of acidity stops falling the process is complete. Again if the ferrous content of these samples shows an undue increase stop the process immediately.

After the chelating rinse, fill the boiler with a 10% sodium carbonate solution and boil for two hours and then discharge to waste, checking that this is alkaline. Again flush and inspect the boiler internally. Remove any restrictions fitted, and use a hose to flush out any parts where poor circulation may have occurred. The internals are replaced and the boiler searched in the normal manner. The boiler is then stored in the recommended manner until required for use.

Before cleaning it is advisable to analyse samples of the deposits so the most suitable chemicals can be chosen; this is especially important if a high copper content is suspected, as a different procedure will be required.

It may be necessary to cut out one or two tubes for inspection before the final decision to acid clean is made. These tubes should be taken from high heat exchange regions of the boiler.

CHAPTER 12

Evaporators

There are various means of producing fresh water from sea water, such as ion-exchange and electro-dialysis, but for most marine purposes evaporation provides the most economically viable method.

This basically uses heat from some convenient source to boil the sea water. The vapour driven off leaves its dissolved solids behind in the water remaining in the evaporator. This water is usually referred to as brine. The problem then arises that as the quantity of these dissolved solids builds up in the brine, they begin to deposit as scale so reducing the efficiency of the heat transfer process. In addition the increased density leads to foaming at the water level, so giving a greater possibility of the carry-over of water droplets along with the vapour leaving the evaporator thus reducing the purity of the fresh water produced.

By operating the evaporator at low pressures and temperatures, and by maintaining reasonably low densities these problems can be greatly reduced. The level of the brine density is controlled by blowing down; this removes both the salts still in solution together with suspended particles of sludge. In general the density of the brine should not exceed 64 000 ppm, with a normal operating density of around 48 000 ppm. The lower the density the smaller the amount of scale depositing but the greater the amount of blow down required, with consequently a greater consumption of heating steam. In practice this means a compromise must be reached between the reduction in cleaning and the increase in fuel consumption. Deposits of scale can often be removed from the heating surfaces by thermal shock when the evaporator is operating, and by mechanical or chemical means when it is shut down.

Small amounts of suitable chemicals added to the evaporator feed are beneficial in reducing the amount of scale formed, or ensuring that only soft scales are deposited.

The fresh water produced can be used to provide make-up feed for the boilers, or for domestic purposes. In this latter case it is referred to as potable water, and the vapour must be passed through a suitable distilling plant and also be sterilised before use.

The evaporator feed can consist of fresh water obtained from shore supplies, or sea water. A separate feed pump may be used, but in many cases the feed is obtained from a suitable main, such as a sea water cooling line. When chemicals are injected into the sea water cooling lines to protect them from corrosion or fouling, care must be taken that the evaporator does not become contaminated. This is especially important when the production of potable water is involved, and in this case a separate untreated sea water supply line will be required for the evaporator.

Many types of evaporators are used at sea, the variations in design usually being to give a more economical use of the heating medium, and (or) to reduce cleaning, rather than to improve the purity of the water produced. This usually has a density of less than 4 ppm and, in the case of modern combined evaporator-distillers working under vacuum conditions, less than 1 ppm.

The make-up requirements of high pressure water tube boilers often demand densities consistantly below 0·5 ppm and this may be achieved by double evaporation, consisting of two evaporators placed in series, the feed of the second being the water produced by the first. Alternatively, using fresh water feed in a single evaporator can give similar results. In other cases the make up feed is passed through a demineralizer which, unlike the previous methods, also enables the dissolved carbon dioxide content of the make-up water to be reduced.

TYPES OF EVAPORATORS

These can be divided into boiling and flash type evaporators, each of which can take the form of a single or multiple effect plant.

SINGLE EFFECT EVAPORATORS

These form simple compact units suitable for small to medium outputs. Supplying live steam to the heating section can provide a high mean temperature difference enabling a small heating surface to be utilized for a given output. This however is thermodynamically uneconomical, and in many cases use is made of bled steam or waste heat, for example the jacket cooling water from a diesel engine. In general this will increase the size of the plant.

SUBMERGED COIL TYPE

The evaporator shown in Fig. 60 contains a number of horizontal heating coils, each of these being attached to a steam chamber formed in the side of the evaporator shell.

The shell is of fabricated mild steel with a bonded rubber coating to protect it against corrosion, while the heating coils are of aluminium brass. The steam supply pressure is kept higher than the exhaust by making the outlet hole in each coil smaller than the inlet as shown in Fig. 61. Both the holes in the bottom coil are the same size, and the direction of flow is reversed so that all the exhaust from the previous coils passes through it. This ensures that any remaining steam is condensed so that only water leaves through the drain. The usual mountings are fitted which include an automatic feed regulator, and in some cases a special brine ejector. The ejector shown in Fig. 62 uses water from a suitable service, such as the sanitory main, which when passed through the ejector nozzle pumps brine from the evaporator. This continuous blow down enables a constant density to be maintained in the evaporator. Alternatively a small brine pump may be fitted.

A baffle is fitted in the vapour space to reduce the carry over of water from the evaporator.

When potable water is required the vapour must be passed through a suitable distilling and sterilising unit.

For safety requirements the evaporator is treated as a steam boiler and new shells and coils must be tested to 2 x Maximum working pressure, by means of a hydraulic test. Safety valves must be fitted and adjusted to lift at the required pressure. A test must also be carried out for accumulation of pressure, this being

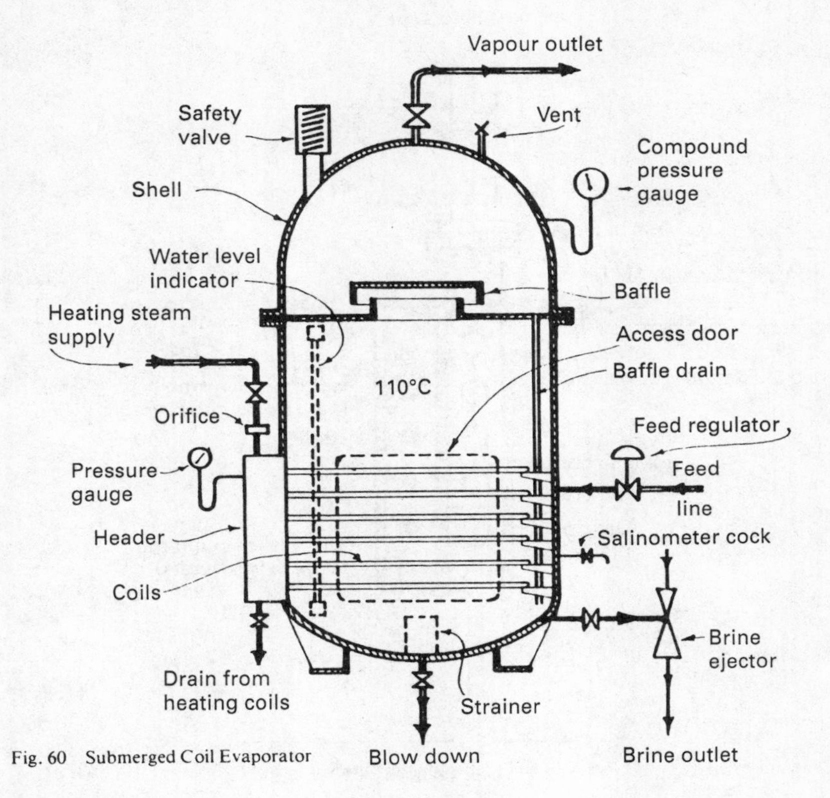

Fig. 60 Submerged Coil Evaporator

done by removing one coil, and then supplying full heating steam to the coils. The pressure in the shell must not now exceed the blow off pressure by more than 10%.

An orifice plate is often fitted to restrict the flow of heating steam and enable a smaller safety valve to be fitted. This orifice must be of non-corrodible metal, and the hole must be parallel for at least 6 mm of its length.

If the shell consists of a single casting or is made of cast iron, the working pressure must not exceed 200 kN/m². Cast iron, bronze, and gun metal must not be used for operating temperatures above 220°C.

OPERATION. The water level should be such as to leave the top coil exposed, as this helps to dry the vapour.

The evaporator should be blown down at regular intervals, if no brine ejector is fitted, to prevent an undue density building up causing excessive scale formation. The density can be measured by drawing off a sample of brine into a salinometer pot and then testing it with a hydrometer.

Blowing down can also be used to remove scale by thermal shock. The procedure for this is to shut the vapour outlet and feed valves, and then using the blow down valve to empty the evaporator. The steam supply valve is now closed and the feed valve opened. The relatively cold feed water entering the shell

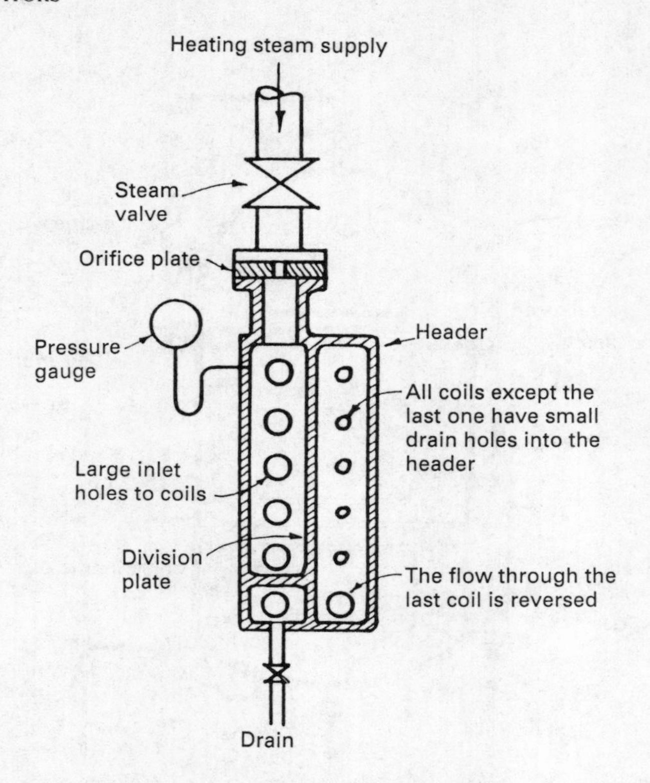

Fig. 61 Evaporator Header

condenses any remaining vapour so forming a partial vacuum which causes sea water to be rapidly drawn in through the blow down valve. The sudden heating followed by sudden cooling tends to crack off any hard scale formed on the coils. When the water level reaches half glass, the blown down valve is closed and the steam and vapour valves opened to return the evaporator to service. When there is no direct sea connection to the blow down the evaporator will have to be refilled with feed in the normal manner.

SUBMERGED TUBE HIGH VACUUM TYPE

This consists of a two part shell, fabricated from mild steel, with its internal surfaces protected against corrosion by a bonded rubber coating. The lower evaporating section contains a vertical tube stack which consists of plain aluminium brass tubes expanded into tube plates at both ends. The upper vapour shell contains the distilling condenser consisting of aluminium brass hairpin tubes expanded into a single tube plate and placed horizontally above a water catchment tray. To reduce carry-over to a minimum, the vapour entering the distilling section has to pass through a mesh type demister. This consists of layers of knitted monel metal wire. Alternatively polypropylene mesh may be fitted,

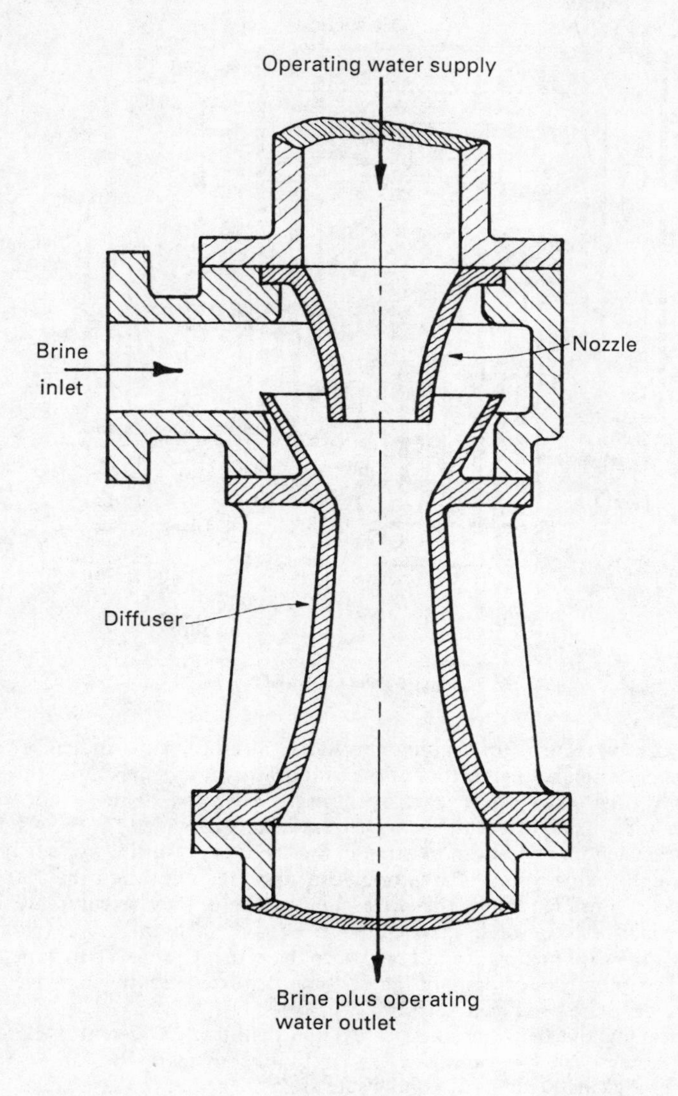

Fig. 62 Brine Ejector

but its use is limited to vapour temperatures below 75 °C. The basic layout of this combined evaporator-distiller is shown in Fig. 63.

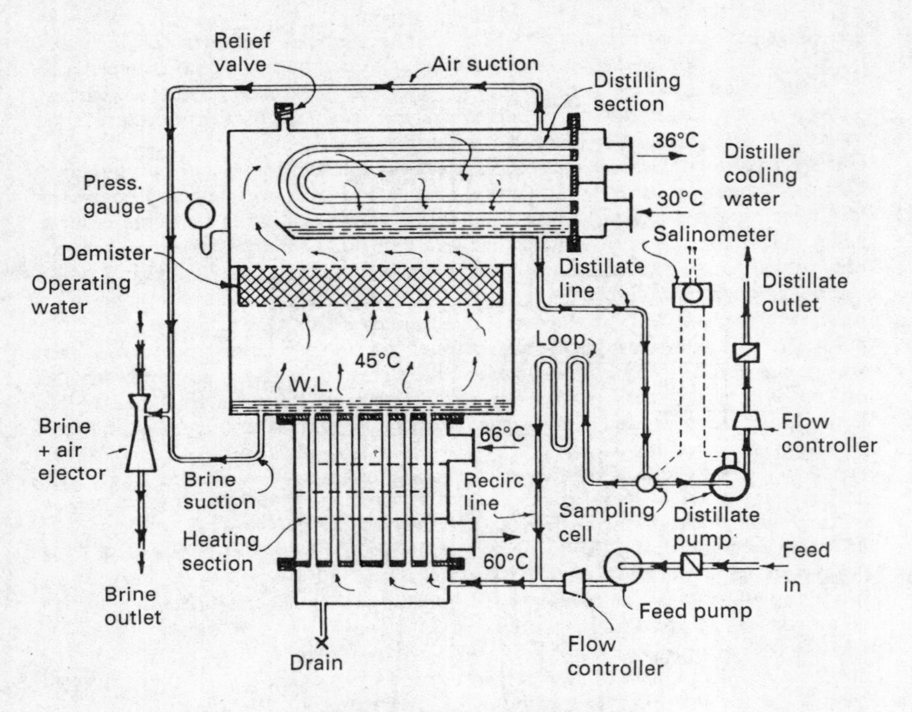

Fig. 63 Combined Evaporator and Distiller

The evaporator feed, after passing through a strainer, flow indicator, and flow controller, enters at the bottom of the evaporator. It then passes up through the vertical heating tubes where its temperature is raised by steam or hot water passing over the outside of the tubes. Sufficient heat is provided for the water to boil under the vacuum conditions existing in the shell, the resulting vapour rising to pass through the demister. The vapour can pass freely through this but any water particles impinge onto the wire mesh, where they accumulate and ultimately coalesce into water droplets large enough to break free, dropping down against the vapour flow, to fall back into the brine. Compared to vane type baffles or flat plate deflectors, this mesh type demister greatly improves the purity of the distillate produced.

Vapour leaving the demister then enters the distilling condenser, where its latent heat is removed by cooling water circulating through the tubes of the distiller. The resulting droplets of condensate are collected in the catchment tray, from where it flows via a salinometer probe to the distillate pump. This probe transmits a signal to the electrical salinometer which measures the density of the distillate. When this is acceptable the distillate pump discharges it through a flow controller and a non-return valve to the storage tanks. If the density is unacceptable the salinometer provides a signal which stops the pump. This allows the unacceptable distillate to pass over the double loop to re-enter the evaporator feed line for re-distillation. As an alternative arrangement the salinometer may be used to operate a series of diverter valves which achieve a similar object.

The brine density is controlled by fitting flow controllers in the feed and distillate lines, these being set to admit 2·75 times as much feed water as the amount of distillate produced, the excess being pumped out by the water operated ejector. This both provides a continuous blow down of brine so as to maintain the density low enough to prevent scale forming, and also removes air and other non-condensible gases released during the evaporation process, from the upper part of the vapour shell. The necessary vacuum for the proper operation of the plant is thus achieved.

About thirty times as much operating water is supplied to the ejector as the amount of brine it removes and this so dilutes the brine that no undue build up of deposits should occur in the discharge lines.

Another factor involved in maintaining the correct brine density is that when newly cleaned the unit can often produce more distillate than its normal rated output. If this is allowed to occur it would reduce the feed to distillate ratio and so lead to an unduly high brine density. The flow controller in the distillate line prevents this by keeping the discharge rate constant, so that any excess distillate returns over the loop to the evaporator, so diluting the brine to its normal density.

Heat for evaporation can be provided by a direct steam supply, by bled steam, or by waste heat such as jacket cooling water from a diesel engine.

The cooling water for the distilling condenser can consist of feed water from a turbine feed system, or of sea water. In the latter case some of the sea water. coolant may be bled off to serve as evaporator feed.

If it is to be used as potable water, then the distillate must be passed through a suitable filter and sterilizing unit, before entering the storage tank.

OPERATION. If the feed supply is interrupted, then in order to avoid an undue build up of density the distillate pump should be stopped until the feed supply has been restored.

The salinometer probe should be cleaned at every available opportunity, while at least once every six months or more frequently if required, the plant should be shut down and cleaned. Any sludge which may have accumulated at the bottom of the shell is washed out and, if necessary, any deposits removed from the heating tubes by acid cleaning. This is done by circulating a 10% solution of hydrochloric or sulphamic acid through the heating section, pumping it in through the feed inlet connection and out through the brine outlet connection in the shell.

FLASH EVAPORATOR

As the majority of scale formation occurs only when the water actually boils, this type of evaporator separates the heating and boiling processes to different sections of the plant, the basic layout of which is shown in Fig. 64.

The incoming sea water feed is first heated to some temperature below its boiling point in tubular heat exchangers. This is done in two stages, in the first formed by the distilling condenser the feed is heated by the outgoing vapour which gives up its latent heat, and in the second stage by steam or hot water. The heated feed water is then released into the flash chamber where the pressure is maintained low enough to ensure that the corresponding saturation temperature

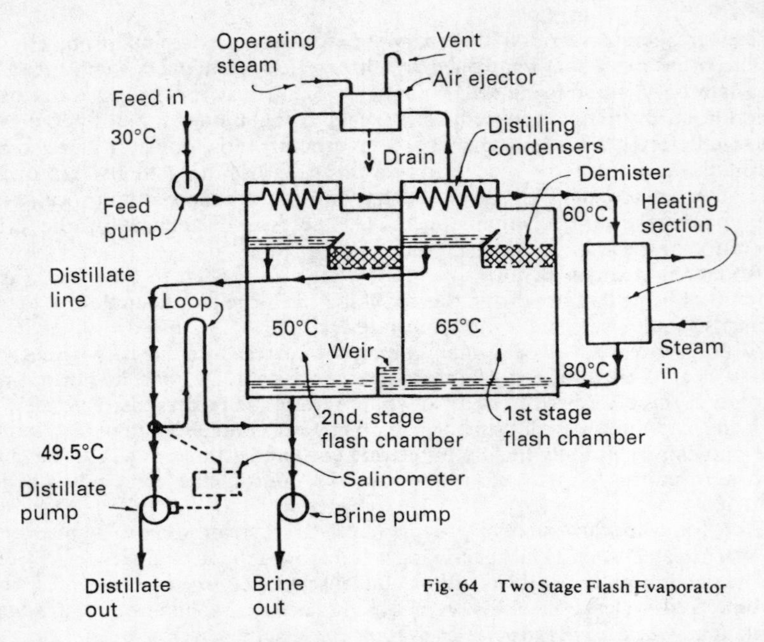

Fig. 64 Two Stage Flash Evaporator

is below that of the incoming hot water. The water cannot remain in this supersaturated state, so some of its mass flashes off into steam, leaving its dissolved solids behind in the water remaining in the flash chamber. The released steam then passes to the distilling condenser, from where the resulting distillate is pumped to the storage tanks via a potable filter and steriliser if necessary. The density of the distillate is measured by a salinometer and if it reaches too high a value, the distillate pump is stopped and the unacceptable water passes over a loop to the brine pump suction from where it is discharged along with the outgoing brine. This is removed from the flash chamber and pumped overboard, being diluted with sea water if necessary to prevent an undue build up of deposits in the discharge line.

By avoiding boiling in the heating sections, scale deposits on the heat exchange surfaces are largely avoided, final water temperatures below 80°C being recommended to keep deposits to a minimum. If necessary chemicals may be added to assist in this, being essential if water temperatures above 80°C are used in the heaters. Any deposits forming in the flash chamber do not interfere with the heat transfer and so do not effect the operation of the plant.

A water operated ejector is used to remove air and other non-condensible gases from the vapour chamber and so maintain the necessary vacuum conditions.

The vapour chamber is fabricated from mild steel, with a bonded rubber coating as a protection against corrosion. Aluminium brass tubes expanded at both ends into tube plates of rolled admiralty brass, form the heat exchange surfaces. The usual mountings for the proper working of the evaporator are fitted.

Automatic control is necessary to ensure satisfactory operation of the plant, as it is sensitive both to changes in sea temperature and to the water level in the flash chamber. Special starting up arrangements will be required as, until flash off commences, the second stage heater has to supply all the necessary heat. This can be done either by supplying additional heating steam, or by restricting the flow of feed into the evaporator.

OPERATION. Salts mainly deposit as sludge in the flash chambers and should be washed out as required when the evaporator is shut down.

MULTIPLE EFFECT EVAPORATION

As an evaporator does no mechanical work its performance cannot be measured in terms of efficiency and instead a performance ratio of distillate output to the heat input, often termed the gain ratio, is used.

To increase the gain ratio the evaporating process may be carried out in a number of pressure stages, or effects. This is achieved by arranging a number of evaporators in series, each being operated at a progressively lower pressure and using the vapour or water from the previous stage to provide the heat for the evaporation process. Both submerged coil and flash evaporators can be arranged in this way.

The larger the number of effects the greater the saving in steam but at the cost of increased complexity, capital cost, and volume occupied by the plant. Thus the law of deminishing returns applies, and where relatively small outputs are involved single stage evaporation is often preferred, as these provide simple compact units. Economy is improved if desired by using bled steam, or waste heat, however for larger outputs economy becomes increasingly important and so warrants the increased cost and complexity of multiple effect plants. For marine use however, except in a few special cases, they are normally limited to double effect flash evaporation plant.

DOUBLE EFFECT FLASH EVAPORATOR

As shown in Fig. 64 the incoming feed is preheated as it passes through the distilling condensers of the successive stages and then raised to 80°C in an external heater. The hot water then enters the first stage chamber where flash off occurs. The released vapour passes through the demister to enter the distilling condenser where it gives its latent heat to the incoming feed. The brine meanwhile flows through to the second vapour chamber which is maintained at a lower pressure than the first, so that further flash off can take place. A weir is fitted to maintain a water seal between the two stages.

Distillate formed in the two vapour condensers is collected in the catchment trays, from where it flows to the distillate pump via a salinometer probe. If the density proves to be too high, the pump stops and the unacceptable water passes over the loop to the brine pump suction where it is discharged with the outgoing brine.

The vapour shells are fabricated from mild steel with a bonded rubber coating although this tends to increase the size of the unit. Although tolerable for single or two stage plant it is not acceptable for greater numbers of stages and a different form of construction is then used. This usually takes the form of a specially fabricated steel structure, protected against corrosion by the application of a special paint. Some form of ejector will be fitted to remove air and other non-condensable gases from the vapour shells. The usual mountings for the proper operation of the plant are also fitted.

Automatic control will again be necessary for efficient working of the plant.

OPERATION. Similar to that required for the single effect plant.

VAPOUR COMPRESSION EVAPORATOR

Another method of achieving steam economy is by means of vapour recompression. This process can given gain ratios in the order of 8 to 1, as compared to 0·9 to 1 for single effect submerged tube, and 1·5 to 1 for double effect flash evaporation. The basic layout of a vapour compression type evaporator is shown in Fig. 65.

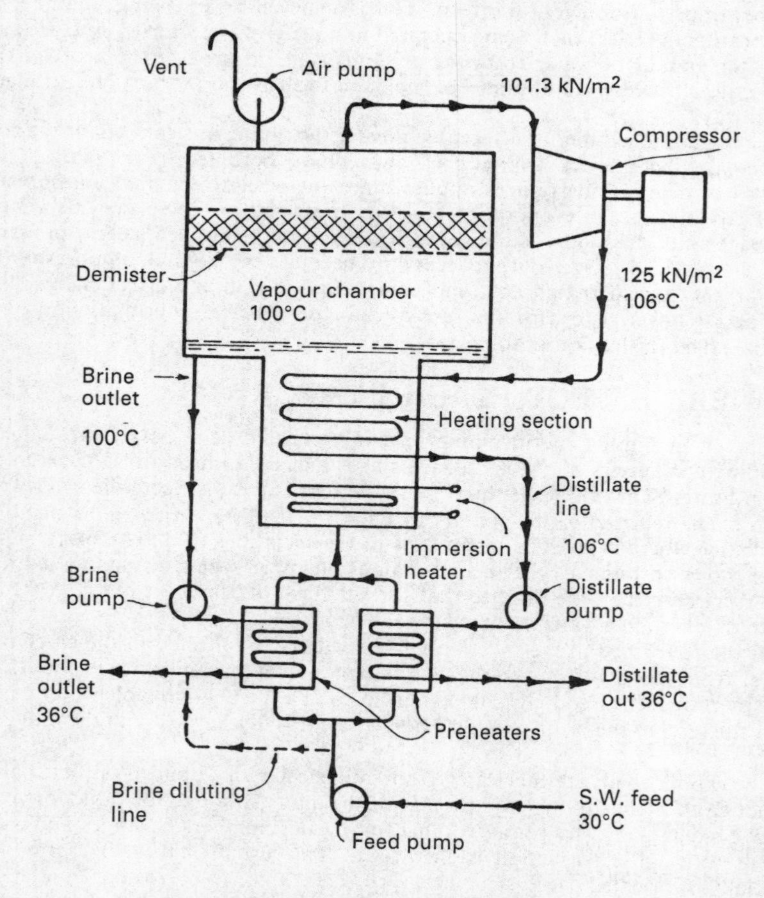

Fig. 65 Vapour Compression Type Evaporator

The vapour is generated in the evaporator shell at atmospheric pressure and then passes through the demister to the compressor suction. Work is now done on the vapour so that it leaves the compressor at a higher pressure and temperature. This compressed vapour passes to the heat exchange section where it gives up its latent heat so raising the temperature of the incoming feed to its boiling point. The vapour produced then repeats this cycle.

The distillate leaving the heat exchange section is pumped to the storage tanks via the usual salinometer probe, so that in the event of the density being too high the salinometer can send a signal to a series of solenoid operated valves causing the unacceptable water to be dumped. Meanwhile brine is pumped out of the vapour shell and discharged overboard. Water bled off from the evaporator feed line is used to dilute this water to prevent deposits from choking the pipes and valves.

To conserve heat the outgoing distillate and brine are passed through heat exchangers, where they give up sensible heat to preheat the incoming sea water feed.

A mechanical or thermo compressor can be used. The latter consists of a steam jet ejector, while the former can be either a high speed centrifugal compressor or a low speed rotary lobe type blower. Provided an effective demister is fitted to prevent carry over of water droplets which would otherwise cause serious erosion of the lobes, the slow speed blower type is to be preferred.

The vapour shell is fabricated from mild steel with a bonded rubber lining. This consists of a 3 mm thick layer of a hard rubber known as ebonite, which is applied to a steel surface which has been shot blasted prior to the application of a bonding compound. The rubber coating is then rolled to squeeze out any air bubbles and then cured by steam heat. This layer provides protection against both the brine and any cleaning acids used. The heat exchange surfaces consist of aluminium brass tubes expanded into rolled Admirality brass tube plates.

For starting up, a separate heat source must be used to provide sufficient vapour to commence the cycle of events; in the plant considered an immersion heater is fitted for this purpose. An air pump is also fitted to remove air from the shell when starting up. This type of plant is normally automated, the programmed start up and shut down procedures being initiated by a single switch. This allows feed to enter the evaporator where the immersion heater raises it to boiling point. As the vapour pressure builds up it causes the air pump which has been removing air from the shell to cut out. The compressor then starts, followed by the brine and distillate pumps. Normal control is carried out to maintain a constant pressure in the vapour shell.

Vapour compression plant when fitted with a mechanical compressor can operate with an electrical input only and so can be used where no steam supply is available.

OPERATION. Due to the fact that evaporation in this type of plant occurs at atmospheric pressure, it is essential that some form of chemical feed treatment be applied. Heat exchange surfaces must be kept clear of deposits and should be cleaned at least once every six months, or more frequently if found to be necessary. Acid cleaning may be required to remove the deposits formed.

PRODUCTION OF FRESH WATER BY REVERSE OSMOSIS

Osmosis depends upon the use of a semi-permeable membrane which allows water molecules to pass freely in either direction, but will not permit the passage of salt molecules.

If two solutions of different densities are separated by such a membrane, the difference in the chemical potentials of the two solutions causes water from the weaker solution to pass through the membrane, tending to reduce the density, and thus the chemical potential, of the stronger solution. However, as the stronger solution gains water its volume increases, and so its level rises to produce a hydrostatic head across the membrane. Eventually the resulting pressure difference reaches a value known as the osmotic pressure where the process stops. If now an external pressure is applied to force the higher level down again, the water is forced back through the membrane as the volumes again equalize. As the salt molecules cannot pass through, the density on this pressurized side again increases, returning to its original value as the levels finally equalize.

It follows that if this applied pressure continues to increase the process will also continue, the density on the pressurized side now rising above its original value as more and more water is forced across the membrane. This reverse process is referred to as reverse osmosis.

Thus if sea water is continuously pumped at high pressure into a vessel containing a suitable membrane, pure water will pass through the membrane and can be drawn off, while the salt unable to pass through remains in a concentrated solution which can then be discharged.

Various types of membrane are in use, some made of hollow, fine fibres of cellulose acetate, some of spirally wound polyamide, while others take the form of a series of plate-type membranes. The final choice of the type used usually

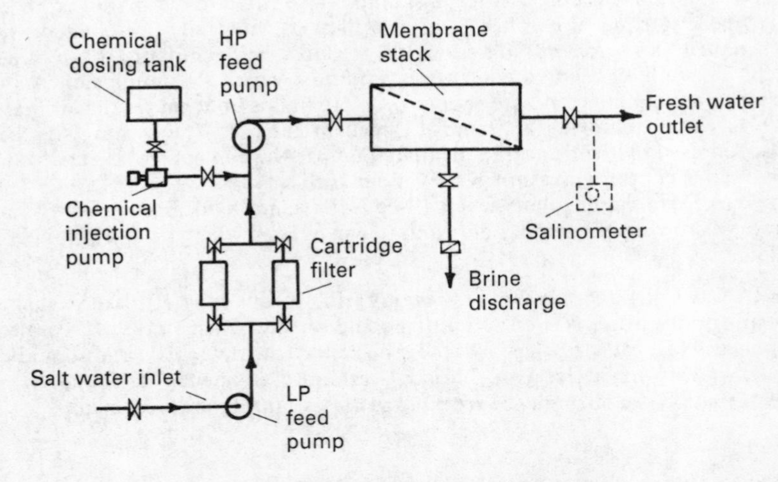

Fig. 66. Single Stage Reverse Osmosis Plant

depends upon the initial water conditions expected, some membranes being more prone to fouling and choking than others.

In order to prevent scale forming on the membranes, continuous dosing of a suitable chemical into the sea water being fed to the unit is required. If potable water is being produced, the chemical must be one approved for this purpose.

Reverse osmosis plants normally operate at temperatures below 40°C and with applied pressures in the order of 6500 kN/m^2.

The basic layout of a single-stage reverse osmosis plant is shown in the form of a block diagram in Fig. 66.

A salinometer is fitted to monitor the output, and if required for potable use the product water must be sterilized.

INDEX